# Laser Beam Mode Selection by Computer Generated Holograms

Victor Soifer
Mikhail Golub

CRC Press
Boca Raton Ann Arbor London Tokyo

**Library of Congress Cataloging-in-Publication Data**

Soifer, V. A. (Viktor Aleksandrovich)
Laser beam mode selection by computer generated holograms / Victor Soifer, Mikhail Golub.
p. cm.
Includes bibliographical references and index.
ISBN 0-8493-2476-9
1. Holography—Data processing. 2. Laser beams—Diffraction. 3. Image processing—Digital techniques. 4. Computer graphics. I. Golub, M. V. (Mikhail Vladimirovich). II. Title.
TA1542.S65 1994
621.36′75—dc20 94-5202
DNLM/DLC CIP
for Library of Congress



Direct all inquiries to CRC Press, Inc., 2000 Corporate Blvd. N.W., Boca Raton, Florida, 33431.


International Standard Book Number 0-8493-2476-9
Library of Congress Card Number 94-5202
Printed in the United States of America 1 2 3 4 5 6 7 8 9 0
Printed on acid-free paper

# PREFACE

Diffractive optics and digital holography are the emerging technologies that give the opportunity to substitute conventional refractive lenses by planar diffractive microrelief etched on some substrate. The role of diffractive optical elements as alternatives for bulk lenses definitely can be compared with the role of electronic circuits and microprocessors as alternatives for electronic valves. At present, diffractive microlenses are under the scope of view in many companies. Familiarity with digital holography and diffractive optics answers the needs of optical engineers, researchers, and optical systems designers.

Two main cores of this book correspond to the two subjects dealt within our work: 1) computer-generated holograms and diffractive optical elements, and 2) selection of coherent modes of fiber and laser beams. The interrelationship between these two aspects provides, in the author's viewpoint, not a mere sum but a certain new approach.

This book may be interesting for scientists, engineers, postgraduates, and students working in the area of advanced optics and fiberoptic systems. Readers of this book should have some background in modern optics, Fourier systems, light propagation, eikonal method as well as in the mathematical techniques of differential equations, theory of series, and numerical methods.

Those who take an interest in digital holography and diffractive optics will find here some materials in nontraditional interpretation. Those who deal with new fiber and laser optical systems, will find here fresh approaches and novel optical elements — modans presented as the components with completely new possibilities. The implementation of diffractive micro-optical elements in fiberoptic systems is indeed a challenging opportunity for advanced engineering. Mathematically-oriented readers will find here some special kind of perturbation theory for orthogonal expansions and a wide possibility for applying different methods of mathematics to fiberoptic and laser systems.

This book is dedicated to the special application of digital holography that is really new for optical science and engineering. This application is in the selection of modes of lasers and optical fibers. This book has three basic lines that are crucial for all other considerations:

- — Complex eikonal method resulting in the strict mathematical definition and description of the modes of laser radiation;
- — Generating function method opening up an opportunity to select arbitrary modes and mode groups;
- — Coding from the complex to the phase-only transfer function that appears important for the generation of diffractive optical elements.

To facilitate the understanding of this book, we would like the reader to be aware of the principle of the material presentation we have chosen. The main line of presentation follows the needs of mode selection. Nevertheless, all the required methods of diffractive optics and digital holography are explained at the proper places of the book. Several methods that are described here are very useful for computer-aided generation of diffractive and holographic optical elements.

The methods and algorithms presented in this book have been tested in real applications. It is a summary of 12 years of scientific activities by the authors on the development of diffractive optical elements. We have used this material in training students, postgraduates, and engineers.

The authors wish to state here that various investigations essential for this book have been conducted at the Image Processing Systems Institute of the Russian Academy of Sciences, Samara. We also wish to express our gratitude for valuable discussions to the postgraduates, students, and faculty members of the Technical Cybernetics Department of the Samara State Aerospace University.

We direct our acknowledgments personally to Iosif N. Sisakyan who participated in conducting the theoretical investigations, and to Sergey V. Karpeev and Gennadiy V. Uvarov who undertook the experimental studies of modans.

In our endeavor to write this book in the English language we have counted on the cooperation of Margarita I. Kotlyar. On completion of this book we take pleasure in expressing our sincere thankfulness for her help.

The manuscript preparation has been conducted with the aid of the publishing departrment headed by Yakov Ye. Takhtarov and with direct participation from Galina V. Yeriomenko, Svetlana N. Khonina, and Andrey E. Tzaregorodtzev who have greatly facilitated the authors' efforts in the final stages.

# INTRODUCTION

Vibrations and wave phenomena are found in various fields of science and engineering including optics, lasers, mechanics, and acoustics. Normal or natural vibrations are distinguished states that are characterized by the common term "modes". Modes can have similar visual representation for such different applications as mechanical vibrations in machines, multimodal beams in laser cavities, and optical fibers.

Each wave-field can be treated as containing the composition of base-functions called modes. The modal approach is based on the decomposition of a wave-field into the orthogonal mode functions that have well-defined physical meaning. Such well known objects as Gauss-Laguerre or Gauss-Hermite modes obey the eigenfunction equation in quadratic-index fibers as well as in free space. Representation of wave-fields by modes is adequate to physical nature and opens good possibilities for detailed, but simple analysis in various engineering applications. Furthermore, the terms "modes of coherent beams" and "multimodal fibers" have already become a part of language in optical and quantum electronics, laser, and fiber optics.

The natural question can arise: whether the modes are mathematical base-functions or they are the physical reality. The authors are convinced that the modes are the physical objects but this must be proved experimentally. Thus, it should be made possible to select and measure any required modes to ensure their physical existence.

It is well known that the selection of longitudinal modes can be performed using the diffraction grating resolving light in terms of wavelengths. The aim of the authors' studies was to develop a novel spectral device capable of measuring the spectrum of transverse modes. Detailed analysis has shown that resolving light in terms of transverse modes involves very complicated optical elements. Digital holography and diffractive optics technology open up an opportunity to develop proper optical elements called "modans".

Current application of digital holography is specific as it makes use of multimodal beams rather than of spherical or plane waves. The general idea here is to influence selectively each transverse mode of coherent beam instead of influencing the whole beam. This idea seems rather specific from the first approach. Nevertheless, as soon as we obtain mode-selective optical elements-modans, we can think about new types of fiberoptic communications with modal multiplexing. Moreover, fiberoptic sensors with mode selection have already been tested in our laboratories and demonstrated very high sensitivity even in the first experiments. Modans as new optical elements furnish an opportunity to measure modal power distribution at fiber's or laser's output. Conventional Fourier lenses could perform such a work only in combination with a digital processor. So new elements-modans perform some optical processing of the beam, not only simple focusing. Modans also create physical standards of single modal beams with different mode numbers that could not be excited by conventional optical methods. This can be treated in such a way that modans build the connecting bridge between the computer images of modes and real optical beams. The combination of multibeam modans and large microlens arrays will give completely new opportunities for information processing and control in coherent optical systems.

This book includes traditional material from digital holography and diffractive optics in authors' interpretation as well as novel results for coherent modes' selection. The physical and mathematical definition for modes of coherent radiation is given in Chapter 1. We start with some mathematical formulas right from the beginning of this book. Such an approach perhaps looks too formal for physical book and engineering applications. However, we have thought many times about the problem "What is the definition for modes in lasers and fibers?" Our brief answer is: "eigenfunctions". Thus, common features of modes could be better described by mathematical equations than by any other language.

Further material of this book is more physical. In Chapter 2 we regard the main problems of mode selection in optical waveguides and introduce the term modan (Section 2.1). Multifocal and multibeam operation are the problems solved in Sections 2.2, 2.3, and 2.4. Section 2.5 deals with approaches of digital holography to coding the complex transfer function into the phase-only one. Thus, the book contains some aspects of optics (Sections 1.1 through 1.5) and digital holography (Sections 2.2, 2.3, and 2.5). The main part (Section 2.4, Chapters 3 through 6) describes new holographic optical elements-modans. Physical approaches (Chapters 1 through 3) are combined with mathematical theory of approximation (Chapter 4) useful for the analysis of sampling effects due to computer generation of continuous-varying transfer functions of optical elements. All the main physical parameters of modan elements have been theoretically evaluated in Chapter 4, including diffraction efficiency and performance accuracy. Experimental studies and their results have been discussed in Chapters 5 and 6.

## TABLE OF CONTENTS

# Chapter 1

# MODES OF COHERENT BEAMS

## 1.1. MODAL BEAMS IN SCALAR APPROXIMATION

A concept of guided modes, or simply modes, naturally arises in laser optics especially for coherent light beams in waveguides with the quantized propagation constant. The Gaussian beams propagating in free space can also be treated as modes.

Let us define modes as the light beams that reproduce themselves during their propagation in a waveguide medium. Modal beams do not become blurred in space and do not change their spatial structure in the proper wave-guide medium. Every mode gets its own attenuation and its own phase delay, which is proportional to the optical path and to the propagation constant.

Thus, the phase delay is continuously accumulated during modes' propagation in the case of a graded-index optical fiber. Furthermore, in this case guided modes reproduce their modal configurations after any path of sufficient length in the case of a graded-index fiber. On the contrary, a guided mode propagating in a cavity is self-reproduced only after every full pass from the second mirror to the first one and back [Anan'ev, 1989; Svelto, 1982].

In this paragraph we consider in details the mode concept. The modes' properties are discussed that are seen as essential with respect to computer generation of optical elements matched to modes' complex amplitudes.

We start with some mathematical formulas right from the beginning of the book as the modes are indeed eigenfunctions of propagation operator. It appeared that there is no good way to define modes as except for the poor attempt to describe equations by words. Thus, we determine modes mathematically in this section and discuss the origin of modes in section 2.1 from a bit philosophic point of view.

Let us set the Cartesian coordinates $(x,y,z)=(\mathbf{x},z)$ in the medium of beam propagation. The two-dimensional vector $\mathbf{x}=(x,y)$ represents the transverse coordinates; $z$ is the longitudinal coordinate on an optical axis. We shall use different designations for the transverse coordinates in several parallel planes $\mathbf{u}=(u,v)$, $\mathbf{x}'=(x',y')$, etc. Guided modes under consideration are thought of as located within the domain $\mathbf{x}\in G$ in the beam cross-section. We shall use the scalar representation of light field and the scalar diffraction theory without any special regard to the polarization effects. Thus, we shall describe monochromatic or quasi-monochromatic field by the complex amplitude $w(\mathbf{x},z)$ corresponding to the wavelength

It must be noted that the modal complex amplitude $\psi_{\mathbf{p}}(\mathbf{x})$ numbered $\mathbf{p}=(p,l)$ is considered on the wavefront surface, and here $\mathbf{x}$ is the coordinate of the wavefront point projection onto the nearest plane perpendicular to the direction of propagation. In the case of the linear medium, we can introduce the linear propagation operator $\hat{P}$ connecting the complex amplitudes $w$ and $w'$ on the two wavefronts separated by some distance:

$$\hat{P}w(\mathbf{x}) = w'(\mathbf{x})$$

The modes $\psi_{\mathbf{p}}(\mathbf{x})$ do not change their transverse configuration and, thus, satisfy the eigenvalue equation

$$\hat{P}\psi_{\mathbf{p}} = \gamma_{\mathbf{p}}\psi_{\mathbf{p}} \tag{1.1}$$

where $\gamma_{\mathbf{p}}$ are the complex eigenvalues. Let us note that the operator $\hat{P}$ and its eigenvalues $\gamma_{\mathbf{p}}$ depend on the distance between the wavefronts under consideration. Thus, the guided modes $\psi_{\mathbf{p}}$ are the eigenfunctions of the propagation operator $\hat{P}$. If the mathematical properties of the normality

$$\hat{P}\,\hat{P}{*} = \hat{P}{*}\,\hat{P} \equiv \hat{Q} \tag{1.2}$$

and compactness [Rudin, 1973] are satisfied, the operator $\hat{P}$ has the complete and countable set of the orthogonal eigenfunctions called modes $\psi_{\mathbf{p}}$ ( * is a symbol of the Hermitian-conjugated operator). Orthogonality property is supposed to be satisfied with some weight of $\rho_0(\mathbf{x}) \geq 0$, $\mathbf{x} \in G$, that depends on a varying tilt of the surface of the modes wavefront. Of course, we can choose the modes $\psi_{\mathbf{p}}$ to be orthonormal

$$\int_G \psi_{\mathbf{p}}(\mathbf{x})\psi^*_{\mathbf{p}'}(\mathbf{x})\rho_0(\mathbf{x})\,d^2\mathbf{x} = \delta_{\mathbf{pp}'} \tag{1.3}$$

by means of the proper normalizing, where * is a symbol of the complex conjugation and

$$\delta_{\mathbf{pp}'} = \delta_{pp'}\delta_{ll'} = \begin{cases} 1, & p = p' \\ 0, & p \neq p' \end{cases}$$

Note that the weight $\rho_0$ will change its form with the changing coordinates $\mathbf{x}$. The powers $\hat{P}^{m_0}$ of the operator $\hat{P}$ correspond to the number $m_0$ of full passes of the beam. The $\hat{P}*$ and $\hat{P}^{m_0}$ operators are commutative with the $\hat{P}$ operator and, thus, have the same eigenfunction $\psi_{\mathbf{p}}$ and eigenvalues, respectively, according to the equations

$$\hat{P}^* \psi_{\mathbf{p}} = \gamma_{\mathbf{p}}^* \psi_{\mathbf{p}}, \quad \hat{P}^{m_0} \psi_{\mathbf{p}} = (\gamma_{\mathbf{p}})^{m_0} \psi_{\mathbf{p}} \tag{1.4}$$

The $\hat{Q}$ operator (1.2) is also commutative with $\hat{P}$, it has real non-negative eigenvalues

$$\lambda_{\mathbf{p}} = \left|\gamma_{\mathbf{p}}\right|^2, \quad \hat{Q}\psi_{\mathbf{p}} = \lambda_{\mathbf{p}}\psi_{\mathbf{p}}, \tag{1.5}$$

and will be called the energy operator.

The numbers $\lambda_{\mathbf{p}}$, Eq.(1.5), characterize the attenuation in modal powers that occurs in one full pass, and the numbers $\arg\gamma_{\mathbf{p}}$ give the phase delay in one pass (arg is a symbol of the argument of the complex number).

In the passive waveguide we have

$$0 \le \gamma_{\mathbf{p}} \le 1.$$

The quantities

$$\alpha_{\mathbf{p}} = 1 - \lambda_{\mathbf{p}} \tag{1.6}$$

characterize the power losses for the $\psi_{\mathbf{p}}$-mode (including the diffraction losses) in one full pass in the waveguide. The mode having the maximum value (and, thus, minimum losses) is called the principal mode. It is usually marked with the zero index $\mathbf{0} = (0,0)$.

$$\lambda_0 = \max_{\mathbf{p}} \lambda_{\mathbf{p}} \tag{1.7}$$

In accordance with Eq.(1.4), the attenuation in the modal power $\psi_{\mathbf{p}}$ for $m_0$ passes is equal to $\left|\gamma_{\mathbf{p}}\right|^{2m_0}$ and the phase delay equals $(m_0 \arg\gamma_{\mathbf{p}})$.

The losses of the **p**-th mode in $m_0$ passes may be determined by the formula

$$\alpha_{\mathbf{p}}^{m_0} = 1 - \lambda_{\mathbf{p}}^{m_0}$$

and are indexed in the same manner with respect to magnitude as the losses in one pass.

Let us consider the mode $\psi_{\mathbf{p}}$ as having the higher order as compared with $\psi_{\mathbf{p}'}$ if the former is characterized by the higher diffraction loss, that is, $\alpha_{\mathbf{p}} > \alpha_{\mathbf{p}'}$ or $\lambda_{\mathbf{p}} < \lambda_{\mathbf{p}'}$. The modes $\psi_{\mathbf{p}}$ and $\psi_{\mathbf{p}'}$, with equal $\lambda_{\mathbf{p}} = \lambda_{\mathbf{p}'}$ for $p \neq p'$, correspond to the degenerate eigenvalues of the Q-operator. They must be indexed in the non-increasing order with respect to the magnitude of the propagation constant, that is, proportional to $\left|\arg\gamma_{\mathbf{p}}\right|$.

Being natural (normal) or eigen-oscillations, the modes of a waveguide may be characterized by the following invariant and optimal properties:

1. Modes are the unique 2-D base-functions that conserve the orthogonality during the guided propagation in their native waveguide media.

2. Single-modal light beam is characterized by less power losses as compared with any other beams that contain modes of the same or higher orders.

3. Modes ensure the best mean-square approximation for the field in the beam's cross-section by the first terms of the orthogonal expansion. Comparison is performed for any other sets of 2-D orthogonal functions.

The property of the modes to reproduce their transverse structure in the process of their propagation was formulated above and can be added to these properties. The mathematical proof of all the foregoing properties is given in Appendix 1.

Having already a concept of the mode that is described by Eq.(1.1), we can represent the whole field $w(\mathbf{x},z)$ in the waveguide medium as the modal superposition:

$$w(\mathbf{x},z) = \sum_{\mathbf{p}} w_{\mathbf{p}} \gamma_{\mathbf{p}}(z) \psi_{\mathbf{p}}(\mathbf{x}), \quad \gamma_{\mathbf{p}}(0) = 1 \tag{1.8}$$

The coefficients $w_{\mathbf{p}}$ of the orthogonal expansion specify the modal composition of the field and are related to it by the equation

$$w_{\mathbf{p}} = \int_G w(\mathbf{x},0) \psi_{\mathbf{p}}(\mathbf{x}) \rho_0(\mathbf{x}) d^2\mathbf{x} \tag{1.9}$$

In conformity with the Parseval theorem of the Fourier expansions, we can write down that the full power of the field $w$ is formed by the powers $|w_{\mathbf{p}}|^2$ of its separate modes:

$$\int_G |w(\mathbf{x},0)|^2 \rho_0(\mathbf{x})\, d^2\mathbf{x} = \sum_{\mathbf{p}} |w_{\mathbf{p}}|^2 \tag{1.10}$$

Let us define modal power distribution as a normalized quantity

$$\frac{|w_{\mathbf{p}}|^2}{\int_G |w(\mathbf{x},0)|^2 \rho_0(\mathbf{x}) d^2\mathbf{x}}$$

Variations of the coefficients $w_{\mathbf{p}}$ versus $\mathbf{p}$ determine the modal composition and depend on the waveguide state.

## 1.2. EXCITATION OF MODES IN OPTICAL FIBERS AND LASER CAVITIES

It is well known that the type of the modal functions depends on the refractive index distribution in an optical fiber or on the configuration and shape of the cavity mirrors. Excitation of modes with specific numbers strongly depends on different factors including the radiation launching conditions, the fiber core diameter, active media and the diffraction losses in a cavity. Traditional equations for modes and their excitation conditions are collected in this section.

When studying the modes in a passive cavity, the operator $\hat{P}$ is some integral operator that describes the light propagation in the cavity from the first mirror to the second one and back, that is, at the distance $2l_0$, where $l_0$ is the length of the cavity (see Fig. 1.1).

It has been shown by Fox and Li and described by [Svelto, 1982] that the complex modal amplitudes $\psi_\mathbf{p}(\mathbf{x}_2)$ for the open cavity, when registered at the output mirror, satisfy an integral equation

$$\int_{D_2} K(\mathbf{x}_2,\mathbf{x}_2')\psi_\mathbf{p}(\mathbf{x}_2')d^2\mathbf{x}_2' = \gamma_\mathbf{p}\psi_\mathbf{p}(\mathbf{x}_2) \tag{1.11}$$

where

$$K(\mathbf{x}_2,\mathbf{x}_2') = \left(\frac{k}{2\pi}\right)^2 \int_{D_1} \frac{\exp[ik(L+L')]}{L\cdot L'}\cdot\frac{(1+\cos\alpha)}{2}\cdot\frac{(1+\cos\alpha')}{2}\cdot \\ \cdot\cos\alpha\cdot\cos\alpha' d^2\mathbf{x}_1;\quad k=\frac{2\pi}{\lambda} \tag{1.12}$$

Here $\mathbf{x}_2,\mathbf{x}_2'$ are the 2-D Cartesian coordinates of the projections of the first or the second mirror's point, respectively, on the plane perpendicular to the optical axis, $D_1, D_2$ are the domains corresponding to the light apertures of the mirrors, $L,\alpha$ are the length and the angle of the tilt to the optical axis of the line segment that connects the point $\mathbf{x}_1$ of the first mirror with the point $\mathbf{x}_2$ of the second one (see Fig. 1.1), $L',\alpha'$ are the similar values as applied to the points marked $\mathbf{x}_1'$ and $\mathbf{x}_2'$. For example, we can assume in a paraxial approximation:

$$L \cong l_0 + \frac{g_1}{2l_0}\mathbf{x}_1^2 + \frac{g_2}{2l_0}\mathbf{x}_2^2 - \frac{1}{l_0}\mathbf{x}_1\mathbf{x}_2,\quad \cos\alpha \cong 1,$$

where

$$g_1 = 1-\frac{l_0}{R_1},\quad g_2 = 1-\frac{l_0}{R_2},$$

$R_1$ and $R_2$ are the radii of curvature of the mirrors' tops.

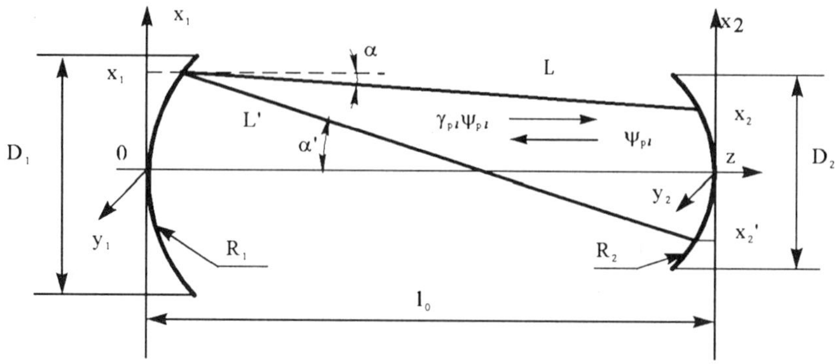


**Figure 1.1.** Modes in the passive open cavity. $l_0$-length of the cavity; $D_1, D_2$-aperture sizes of the mirrors; $R_1$, $R_2$-radii of curvature of the mirrors.

Another case of natural mode representation is given by the beams in a graded-index optical fiber with the refractive index $n(\mathbf{x})$. In this case the operator $\hat{P}$ relates the solution $w(\mathbf{x},z)$ of the Helmholts equation

$$\nabla_{\perp}^2 w(\mathbf{x},z)+\frac{\partial^2 w(\mathbf{x},z)}{\partial z^2}+n^2(\mathbf{x})k^2 w(\mathbf{x},z)=0 \tag{1.13}$$

to the initial values $w|_{z=0}=w(\mathbf{x},0)$ of the complex amplitude, where $\nabla_{\perp}=(\partial/\partial x,\partial/\partial y)$ is the transverse differential Hamiltonian operator (see Fig. 1.2). If the waveguide finite diameter is taken into account, then certain additional boundary conditions appear on the bound of the core.

According to [Unger, 1977], the modes of a graded-index optical fiber have the plane wavefront and obey the equation

$$\nabla_{\perp}^2 \psi_{\mathrm{p}}(\mathbf{x})+\left[k^2 n^2(\mathbf{x})-\beta_{\mathrm{p}}^2\right]\psi_{\mathrm{p}}(\mathbf{x})=0 \tag{1.14}$$

For any given distance $z$ we have

$$w(\mathbf{x},z)=\gamma_{\mathrm{p}}\psi_{\mathrm{p}}(\mathbf{x}) \tag{1.15}$$

$$\gamma_{\mathrm{p}}=\exp(iK_{\mathrm{p}}z) \tag{1.16}$$

$$K_{\mathrm{p}}=\beta_{\mathrm{p}}+iq_{\mathrm{p}}, \tag{1.17}$$

where $\beta_{\mathrm{p}}$ is the propagation constant and $q_{\mathrm{p}}$ is the coefficient of attenuation for the mode $\psi_{\mathrm{p}}$ (see Fig. 1.2). Thus, the modes of graded-index optical fiber satisfy the eigenvalue equation (1.1) for any distance $z$ passed by light. Eigenvalues are specified by Eq.(1.16). It must be noted that Eq.(1.15) describes the modal self-reproduction that occurs with a constant scale along the Cartesian coordinates $\mathbf{x}$ .

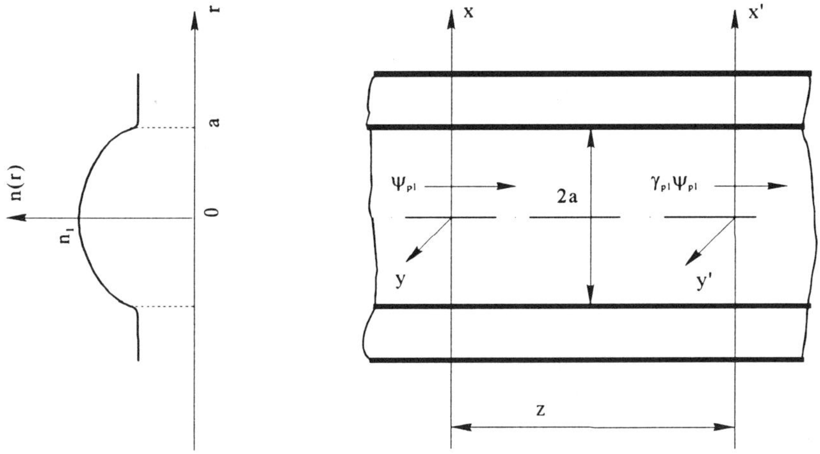


**Figure 1.2.** Modes in graded-index optical fiber $r = \sqrt{x^2 + y^2} \leq a$.

## 1.3. COMPLEX EIKONAL METHOD

One should bear in mind that propagation of a modal beam in free space is a purely diffractive phenomenon and cannot be described by the traditional eikonal function. At the same time, [Felsen, 1973; Solimeno, 1986] have successfully used a complex eikonal method for the Gaussian beams description. The complex eikonal also describes the light fields that emerge inside the caustic and under the condition of the full internal reflection. [Felsen, 1973] have demonstrated that the principal mode of the Gaussian beam can be calculated by the complex eikonal method under paraxial approximation. [Landesman, 1988; Nemoto, 1990] have found such solutions of the Helmholts equation that generalize the Gaussian beams.

In this section we investigate such modal solutions of the Helmholts equation that have common wavefronts of the preset shape. The complex eikonal as well as the complex transport equation are used as a basis for constructing modes of higher orders in the curvilinear coordinates. Such an approach gives, in particular, the non-paraxial beams whose paraxial approximation coincides with the Gauss-Hermite and Gauss-Laguerre modes. The possibility has been shown of building up modes generalizing the Gaussian modes.

A method of the complex eikonal directly generalizes the ray-tracing approximation [Born, 1970] for the medium with an arbitrary distribution of the refractive index $n = n(\mathbf{r})$, in particular, for the uniform medium with $n = n_0 \equiv const$.

Let the required wavefront be chosen from some set generated by the real function $S(\mathbf{r})$ of the 3-D coordinates $\mathbf{r}$ according to the equation $S = const$.

Following [Solimeno, 1986], the additional function $B(\mathbf{r})$ has been introduced so that the equation $B(\mathbf{r})$=const should determine the surfaces of equal intensity. Those surfaces are similar to a hyperboloid of the Gaussian beam (see Fig. 1.3). Complex eikonal is introduced by the equation

$$V(\mathbf{r}) = S(\mathbf{r}) + iB(\mathbf{r}) \tag{1.18}$$

It should be noted that each of the functions $V(\mathbf{r})$, $S(\mathbf{r})$, $B(\mathbf{r})$ does not depend both on the wavelength λ and on the mode number. The complex amplitude under the complex eikonal approach is represented in the same form as in geometrical optics

$$W(\mathbf{r}) = A_0(\mathbf{r})F(\mathbf{r})\exp[ikV(\mathbf{r})], \quad k = 2\pi / \lambda \tag{1.19}$$

Here $A_0(\mathbf{r})$ is the principal term of the asymptotic series expanded in powers of $1/k$ as described by [Born, 1970], and $F(\mathbf{r})$ is some multiplicative component introduced in this work. The component takes into account the remaining terms of the asymptotic series and can be represented in the form:

$$F(\mathbf{r}) = 1 + O(1/k), \quad \nabla F(\mathbf{r}) = O(1/k), \tag{1.20}$$

where quantity $O(1/k)$ is the same order infinitesimal as $1/k$, upon $\lambda \to 0$, $\nabla$ is the operator of the 3-D gradient with respect to $\mathbf{r}$. The imaginary part $B$ of the eikonal V corresponds to the attenuated variations of the field amplitude W by the rule $\exp(-kB)$ taking place due to beam's diffractive blurring. It seems essential that the functions $A_0, F$ as well as all the involved components are supposed to be slowly varying as compared with the factor being fast-oscillating at *exp(ikS)*. The substitution of Eq.(1.19) into the Helmholts equation

$$\nabla^2 W + k^2 n^2 W = 0$$

yields the following equation:

$$\begin{gathered} \nabla^2(A_0F) + 2ikA_0\nabla F\, \nabla V + ikF\left[2\nabla A_0\, \nabla V + A_0\nabla^2 V\right] + \\ k^2 A_0 F\left[n^2 - (\nabla V)^2\right] = 0 \end{gathered} \tag{1.21}$$

By virtue of Eq.(1.20), first two terms of Eq.(1.21) have the order of $O(k^0)$, the second term - the order of $O(k^1)$, and the last two terms have the order of $O(k^2)$. Let us require the function V to satisfy the complex eikonal equation:

$$(\nabla V)^2 = n^2 \tag{1.22}$$

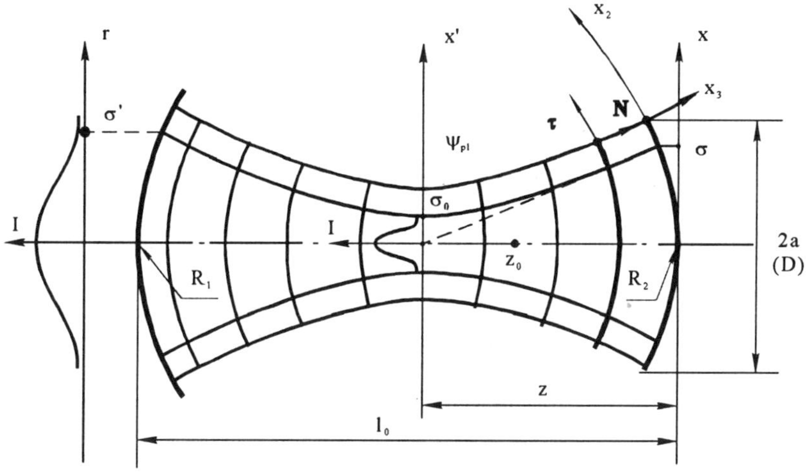


**Figure 1.3.** Gaussian beams in free space. $\sigma_0$ - radius of the waist of principal mode, $\sigma, \sigma'$ - radii of principle mode on mirror's surfaces, $2z_0$- confocal parameter, $I$- gaussian beam intensity, $l_0$- length of the cavity, $R_1$, $R_2$- radii of curvature of mirrors, $2a$ - size of output mirror aperture.

and principal term $A_0$ of an asymptotic expansion to satisfy the transport equation

$$2\nabla A_0\ \nabla V + A_0 \nabla^2 V = 0 \tag{1.23}$$

Equations (1.22) and (1.23) have the same conventional form as in geometrical optics. For the complex eikonal V, the magnitude $A_0$ turns out to be different. Provided Eqs.(1.22) and (1.23), we obtain that Eq.(1.21) for F takes the form:

$$\nabla^2 (A_0 F) + 2ikA_0 \nabla F\ \nabla V = 0 \tag{1.24}$$

This is a generalization on the case of complex eikonal of Eq.(2.6.10), Chapter 2, of the monographic work by [Solimeno, 1986]. The equation (1.24) characterizes diffractive blurring of the light beam in the process of propagation.

Equations (1.22) and (1.23) still have the same forms as in geometrical optics with the real eikonal $V$. Taking into account the representation (1.18) of the complex eikonal gives us specific equations that involve the beam diffractive blurring:

$$(\nabla S)^2 - (\nabla B)^2 = n^2 \tag{1.25}$$

$$\nabla S \cdot \nabla B = 0 \tag{1.26}$$

$$2\,\nabla \ln A_0\ \nabla B + \nabla^2 B - 2i\,\nabla \ln A_0 \cdot \nabla S - i\,\nabla^2 S = 0 \tag{1.27}$$

$$\nabla^2 (A_0 F) - 2kA_0 \nabla F\ \nabla B + 2ikA_0 \nabla F\ \nabla S = 0 \tag{1.28}$$

Let us note that, unlike the corresponding equations proposed by [Solimeno , 1986], herein we did not separate the real and imaginary parts of the value $\ln A_0$, since this appears to be convenient for further calculations. Equations (1.25) through (1.28) are seen to show mutual relations between the phase and amplitude of the beam. Let us call the surfaces $S(\mathbf{r}) = const$ the wavefronts $S$. They are different from the wavefronts of geometrical optics. The set of curves orthogonal to the wavefronts are different from the straight lines of geometrical optics rays (even in free-space) and may be called as generalized rays. The generalized rays have in each point the unique tangential vector

$$\mathbf{N} = \frac{\nabla S}{|\nabla S|} \quad ,$$

collinear to the gradient of the S function. The direction of the quickest amplitude attenuation in each point has the unit vector

$$\tau = \frac{\nabla B}{|\nabla B|},$$

collinear to the gradient of $B$ function and orthogonal to the **N** vector (by virtue of Eq.(1.26)) and, thus, lying in the plane tangential to the wavefronts of S.

The field described by Eq.(1.19) and characterized by the slowly varying function $A_0 F$ has the constant-phase surface $kS + \arg(A_0 F) = const$ being different from the wavefront $S(\mathbf{r}) = const$ and approaching it upon $k \to \infty$.

It is essential that the complex eikonal method, as well as the conventional eikonal method, enables us to replace the light beam in the vicinity of any $\mathbf{r}$ -point by the local plane wave. Actually, let us expand the slowly varying functions $S(\mathbf{r}')$ and $B(\mathbf{r}')$ into a 3-D series, in powers of ($\mathbf{r}'$- $\mathbf{r}$) up to the linear terms. Upon taking into account Eq.(1.19) and the condition $n = const$, we obtain the formula:

$$W(\mathbf{r}') = A_0 F \exp(ikS - kB) \exp[-kn\alpha\tau(\mathbf{r}' - \mathbf{r})] \exp[ikn\beta\mathbf{N}(\mathbf{r}' - \mathbf{r})] \tag{1.29}$$

where

$$\alpha = \frac{1}{n}|\nabla B|,\ \beta = \frac{1}{n}|\nabla S|$$

Equation (1.29) characterizes a fading plane wave near the point $\mathbf{r}(\mathbf{r}' \cong \mathbf{r})$ propagating with the direction **N** and having transverse attenuation $\tau$ along the wavefront. Let us note, that such a plane wave is not a geometrical-optics one, since the module $kn\beta$ of its wave vector

$$\mathbf{k} = kn\beta\mathbf{N}$$

is not equal to the wavenumber $kn$ in the media. Here, instead of geometrical-optics equality we have $\beta \neq 1$, and by virtue of Eq.(1.25), the value β together with the function $\alpha$ satisfies the equation

$$\beta^2(\mathbf{r}) - \alpha^2(\mathbf{r}) = 1 \tag{1.30}$$

It is important to underline that the phase of the transversely fading plane wave (1.29) changes with coordinates in a linear way

$$kS = kn\beta\mathbf{N}\mathbf{r}' \tag{1.31}$$

When passing to the conventional geometrical optics, we achieve $\beta \to 1$ and $kS = kn\mathbf{N}\mathbf{r}'$.

## 1.4. AMPLITUDE-AND-PHASE RELATIONS FOR FREE-SPACE PROPAGATION OF MODE BEAMS

### A. Free-space modes in generalized coordinates

[Haus, 1984; Marcuse, 1982] have described two well-known types of the Gaussian beams: the Gauss-Laguerre and the Gauss-Hermite beams specified only in a paraxial approximation that corresponds to the change from the Helmholts equation (1.13) to a parabolic equation.

In this section we formulate a more general class of the beams having much more general aspherical wavefronts but nevertheless reproducing themselves during propagation.

Being aimed at mode selection with the help of optical elements - modans, one must take into account the change of mode's size during their propagation in free space between the modan and the fiber or the laser. For example, if the beam diverges at the output end of the fiber with high numerical aperture having the plane wavefront at the end, it acquires the complex aspherical wavefront in the process of propagation in free space. Similar beams propagate within and out of the open cavities.

The beams of particular interest are those that change only a scale and a wavefront form not the field distribution along wavefronts during their propagation. Such beams so far reproduce themselves. The beam that can reproduce itself enables us to deal with the same modes on the modan surface as in a fiber waveguide. At the same time, it is the mode ability to reproduce their cross-section structure upon propagating in a waveguide medium that can be characterized as a fundamental property of modes. For example, in optical fibers, there occurs continuous repetition of the spatial structure along the longitudinal z-axis, see Eqs.(1.15) and (1.16). In laser cavities the cross-structure repetition is only with respect to the scale and is periodical and discrete in behaviour along the longitudinal coordinate axis. The Gauss-Hermite and Gauss-Laguerre modes that reproduce their structure in free space along the wavefront with respect to a scale are well known. However, we can say that they actually repeat the structure if normalized coordinates are used. The Gaussian modes are the solutions of Eqs.(1.1) and (1.14) in the paraxial approximation and are known to have the parameters σ and R that represent the size of the principle mode for the $\exp(-2)\cong 0.14$ level of the intensity fall and the wavefront curvature radius, respectively. The Gaussian modes in the graded-index waveguide hold the constant width $\sigma \equiv \sigma_0$, have a plane wavefront ($R=\infty$) and continuously reproduce themselves in accordance with Eq.(1.15) under propagation (see Fig. 1.4). The Gaussian modes, when considered in the passive cavity or in free space, change their width σ=σ(z) and wavefront curvature (radius) $R=R(z)$ (see Fig. 1.3). However, the complex amplitude of the Gaussian mode taken on the wavefronts can be considered as reproducing itself up to varying scale of coordinates σ(z), that is, for the modes specified in a novel normalized coordinates

$$x_1 = \frac{x}{\sigma(z)}, \quad x_2 = \frac{y}{\sigma(z)}, \quad x_3 = z, \tag{1.32}$$

with the scale changing along the z-axis. Thus, in free space the Gaussian beams do not reproduce themselves in the form of simple repetition of their cross-section structure, as is the case for the graded-index fibers, but display the said property only upon introducing the special scaled coordinates. The problem of wavefronts formation with predetermined intensity distribution have discussed in [Golub, 1991a]. The modal light beams in free space capable of reproducing themselves are also required in such task. Here, it turns out to be essential that all of the beams from the complete set must have the common wavefront at a certain location in free space.

Free space modes are built below on the basis of the required set of the wavefronts corresponding to the arbitrary orthogonal coordinates $(x_1, x_2, x_3)$ that generalize the normalized coordinates (1.32). The "longitudinal" coordinate $x_3$ corresponds to the direction of the beam propagation and the "transverse" coordinates $(x_1, x_2)$ characterize position of the point, on the surface $x_3 = const$. The equation aimed at finding the modes that continuously reproduce themselves, can be written in curvilinear coordinates $(x_1, x_2, x_3)$ in the form of eigenvalues equation (1.2). In the present formulation of the problem, Eq.(1.2) on the wavefront $x_3 = const$ takes the form:

$$w_{\mathrm{p}}(x_1, x_2, x_3) = \psi_{\mathrm{p}}(x_1, x_2) \cdot \gamma_{\mathrm{p}}(x_3); \quad \gamma_{\mathrm{p}}(0) = 1 \tag{1.33}$$

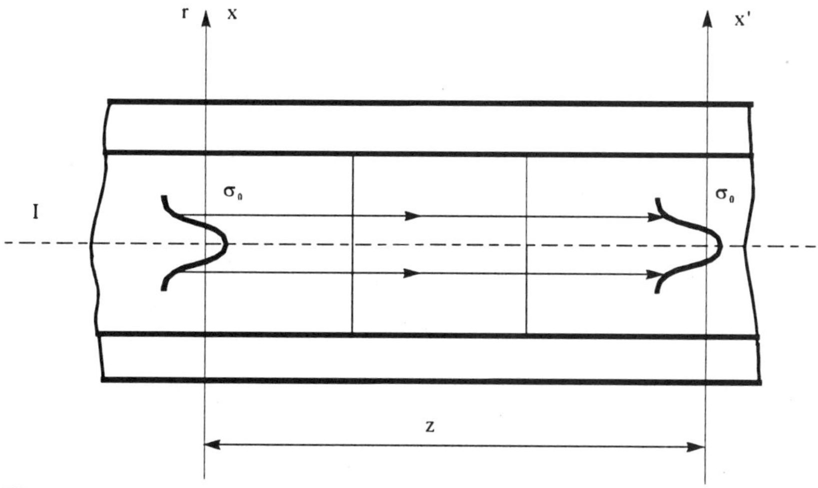


**Figure 1.4.** Gaussian modes in fiber. $\sigma_0$-radius of principle mode; z-distance of propagation .

where $\mathbf{p}=(p,l)$ is the number of the mode, $\psi_\mathbf{p}$ is the complex amplitude of the mode on the original wavefront corresponding to the $x_3=0$ coordinate value, and $w_\mathbf{p}$ is the mode complex amplitude on some output wavefront with coordinate $x_3=const$. The modes' complex amplitude when taken in an arbitrary point, depends both on Eq.(1.33) and on the wavefront form that determines the phase. Thus, the modes with the preset wavefront have rather specific correlation between the amplitude and the phase; the correlation is stable due to the modes' property to reproduce themselves continuously.

**B. Equations for the modes with required common wavefront**

Let us find the modes with the preset set of the wavefronts $S(\mathbf{r})=const$ in the uniform medium using the complex eikonal method. For this we change over to the orthogonal curvilinear coordinates $(x_1,x_2,x_3)$ related to the Cartesian coordinates *(x, y, z)* by the relationships $\mathbf{r}=\mathbf{r}(x_1,x_2,x_3)$ or:

$$\begin{cases} x=x(x_1,x_2,x_3) \\ y=y(x_1,x_2,x_3) \\ z=z(x_1,x_2,x_3) \end{cases}$$

and chosen so that the coordinate planes $x_3=const$ correspond to the wavefronts $S(\mathbf{r})=const$ and the coordinate planes $x_3=const$ correspond the definite intensity fall. Therefore, we can write:

$$S=S(x_3), \quad B=B(x_1) \tag{1.34}$$

In this case the curvilinear lines of the longitudinal coordinate $x_3$ are perpendicular to the wavefronts $S = const$ and they characterize the propagation of the modal beam. At the same time, coordinate lines along $x_1$ characterize the direction of the intensity fall in the beam cross-section.

Comparison of Eqs (1.33), (1.34) and (1.19), (1.18) enables us to write down

$$W(x_1,x_2,x_3) = \psi_{\mathbf{p}}(x_1,x_2)\gamma_{\mathbf{p}}(x_3)\exp[ikS(x_3)] \tag{1.35}$$

$$\psi_{\mathbf{p}}(x_1,x_2)\gamma_{\mathbf{p}}(x_3) = A_0(x_1,x_2,x_3)\,\mathrm{F}(x_1,x_2,x_3)\exp[-ikB(x_1)] \tag{1.36}$$

where $\psi_{\mathbf{p}}$ is the mode numbered as $\mathbf{p}$, $\gamma_{\mathbf{p}}$ is the corresponding eigenvalue in Eq.(1.2), and the $S$-related factor characterizes the wavefront form being common for all modes. Thanks to the separation of the transverse variables $(x_1,x_2)$ from the longitudinal one $x_3$ in Eqs.(1.35) and (1.36), the functions F and $A_0$ are also to be separable by the variables, that is being written down in the form

$$A_0(x_1,x_2,x_3) = A_0(x_1,x_2)\gamma_0(x_3) \tag{1.37}$$

$$\mathrm{F}(x_1,x_2,x_3) = \mathrm{F}_{\mathbf{p}}(x_1,x_2) f_{\mathbf{p}}(x_3) \tag{1.38}$$

where $A_0(x_1,x_2)$, $\mathrm{F}_{\mathbf{p}}(x_1,x_2)$, $\gamma_0(x_3)$, $f_{\mathbf{p}}(x_3)$ - are the new functions introduces here.

Being the principal at $k \to \infty$, the function $A_0(x_1,x_2)$ can be interpreted as one that corresponds to the principal mode. Therefore, from Eqs.(1.36) and (1.20) we can obtain:

$$\psi_0(x_1,x_2) = A_0(x_1,x_2)\exp[-kB(x_1)] \tag{1.39}$$

Here, the F-related factor specified in (1.20) and applied to (1.36), makes the difference of higher order modes from the principal one. From Eqs.(1.36) through (1.39) we derive:

$$\psi_{\mathbf{p}}(x_1,x_2) = \psi_0(x_1,x_2)\cdot \mathrm{F}_{\mathbf{p}}(x_1,x_2) \tag{1.40}$$

$$\gamma_{\mathbf{p}}(x_3) = \gamma_0(x_3)\cdot f_{\mathbf{p}}(x_3) \tag{1.41}$$

Let us re-write Eqs.(1.25) through (1.28) in curvilinear orthogonal coordinates using proper formulae from [Korn, 1968] for the operators $\nabla$ and $\nabla^2$. Thanks to the coordinate separation in representations (1.37) and (1.38) we obtain

$$\left(\frac{1}{H_3}S'\right)^2 - \left(\frac{1}{H_1}B'\right)^2 = n^2 \tag{1.42}$$

$$2\rho_1 B'\frac{\partial \ln A_0}{\partial x_1}+\frac{\partial}{\partial x_1}(\rho_1 B')-2i\rho_3 S'\frac{\partial \ln\gamma_0}{\partial x_3}-i\frac{\partial}{\partial x_3}(\rho_3 S')=0 \quad (1.43)$$

$$\frac{1}{A_0 \mathrm{F_p}}\frac{\partial}{\partial x_1}\left(\rho_1\frac{\partial(A_0 \mathrm{F_p})}{\partial x_1}\right)-2k\rho_1 B'\frac{1}{\mathrm{F_p}}\frac{\partial \mathrm{F_p}}{\partial x_1}+\frac{1}{A_0 \mathrm{F_p}}\frac{\partial}{\partial x_2}\left(\rho_2\frac{\partial(A_0 \mathrm{F_p})}{\partial x_2}\right)+$$
$$\frac{1}{\gamma_0 f_p}\frac{\partial}{\partial x_3}\left(\rho_3\frac{\partial(\gamma_0 f_p)}{\partial x_3}\right)+2ik\rho_3\frac{1}{f_p}\frac{\partial f_p}{\partial x_3}=0 \quad (1.44)$$

$$\rho_1=\frac{H_2H_3}{H_1},\quad \rho_2=\frac{H_1H_3}{H_2},\quad \rho_3=\frac{H_1H_2}{H_3} \quad (1.45)$$

where $H_1, H_2, H_3$ are the Lame coefficients in the Cartesian system and $B', S'$ designate the derivatives of the functions $B$ and $S$ (1.34) with respect to the variables $x_1$ and $x_3$, respectively. Equation (1.43) shows that the function $A_0$ can be chosen as not depending on $x_2$. After having found $S, B, A_0, \gamma_0, \mathrm{F_p}, f_p$ functions from Eqs.(1.42) through (1.44), we can write down the modes (1.39) and (1.40) and the eigenvalues (1.41).

The case of the paraxial beams propagating in the proximity of the z-axis is of particular interest. The formal conditions for this are

$$\frac{|x|}{z}<<1,\quad \frac{|y|}{z}<<1, \quad (1.46)$$

For such beams we can set

$$A_0\cong const \quad (1.47)$$

$$\left|\frac{\partial}{\partial x_3}\left(\rho_3\frac{\partial(\gamma_0 f_p)}{\partial x_3}\right)\right|<<k\rho_3\left|S'\gamma_0\frac{\partial f_p}{\partial x_3}\right| \quad (1.48)$$

and in place of Eq.(1.43) and (1.44) write down the paraxial equations in the form

$$\frac{\partial}{\partial x_1}(\rho_1 B')-2i\rho_3 S'\frac{\partial \ln\gamma_0}{\partial x_3}-i\frac{\partial}{\partial x_3}(\rho_3 S')=0 \quad (1.49)$$

$$\frac{1}{F_p}\frac{\partial}{\partial x_1}\left(\rho_1\frac{\partial F_p}{\partial x_1}\right)-2k\rho_1 B'\frac{1}{F_p}\frac{\partial F_p}{\partial x_1}+$$
$$+\frac{1}{F_p}\frac{\partial}{\partial x_2}\left(\rho_2\frac{\partial F_p}{\partial x_2}\right)+2ik\rho_3 S'\frac{1}{f_p}\frac{\partial f_p}{\partial x_3}=0 \quad (1.50)$$

**C. One-dimensional modal beams**

Let us consider the 1-D beams that have $x_2$-independent complex amplitudes that are generated by the cylindrical-type coordinates:

$$\begin{cases} x = x(x_1, x_3) \\ y = x_2 \\ z = z(x_1, x_3) \end{cases} \tag{1.51}$$

satisfying Cauchy-Rimann - type conditions [Korn, 1968]

$$\begin{cases} \dfrac{\partial x}{\partial x_1} = \dfrac{\partial z}{\partial x_3}, \\ \dfrac{\partial x}{\partial x_3} = -\dfrac{\partial z}{\partial x_1} \end{cases} \tag{1.52}$$

For the beams in the form (1.51) we have

$$H_1^2 = H_3^2 = \left(\frac{\partial x}{\partial x_1}\right)^2 + \left(\frac{\partial z}{\partial x_1}\right)^2; \quad H_2^2 = 1 \tag{1.53}$$

$$\rho_1 = \rho_3 = 1; \quad \rho_2 = H_1^2,$$

where $\rho_2$ is $x_2$-independent.

The eikonal equation for one-dimensional modes is as follows

$$[S'(x_3)]^2 - [B'(x_1)]^2 = H_1^2 \tag{1.54}$$

and it imposes severe constraints on the potential transformation of the coordinates (1.51). Since the variable $H_1^2$ is supposed in (1.54) to be a sum of the functions of $x_1$ and the function of $x_3$, it obeys the equation

$$\frac{\partial^2 H_1^2}{\partial x_1 \partial x_3} = 0 \tag{1.55}$$

We found it convenient to write the transformation (1.51) in the complex plane

$$X = x + iz, \quad U = x_1 + ix_3$$

By virtue of the (1.52), the complex-valued function $X = X(U)$ obeys the conditions of Cauchy-Rimann quoted by [Korn, 1968]. Thus $X(U)$ is the analytical function and the transformation $X(U)$ is a conformal one $(H_1 \neq 0)$. In this case

$$H_1^2 = |X'(U)|^2 = X' \cdot X'^*,$$

where $X'$ is the derivative of $X$ with respect to the complex variable $U$. Equation (1.55) can be rewritten in the form

$$\frac{\partial^2}{\partial x_1 \partial x_3}(X' \cdot X'^*) = i(X'^* \cdot X''' - X' \cdot X'''^*) = 0$$

or

$$\frac{X'}{X'''} = \frac{X'^*}{X'''^*} \tag{1.56}$$

The condition (1.56) means that the function $X'/X'''$ is a real-valued one. Thus, the analytical function $X'(U)$ satisfies the equation

$$(X')'' + \omega_0^2(U)\, X' = 0 \tag{1.57}$$

with the real coefficient $\omega_0^2$. Equation (1.57) is the condition on the class of the allowable coordinate transformations. The entire class of the allowable coordinate transformations for the cylindrical wavefronts can be generated by varying the smooth real function $\omega_0(U)$ in the course of solving Eq.(1.57). Equation (1.57) ensures obtaining such a Lame-coefficient $H_1^2$ that can be represented as a sum of the functions of $x_1$ and the function of $x_3$. To find the eikonals $B$ and $S$ from $H_1^2$ let us use the arbitrariness in the choice of the functions $\theta(x_1, x_3) > 1$ and $\chi(x_1, x_3)$ in the identity

$$H_1^2 = \left(\frac{\theta^2}{\theta^2 - 1} H_1^2 + \chi\right) - \left(\frac{1}{\theta^2 - 1} H_1^2 + \chi\right)$$

and choose them so that the first bracket becomes $x_1$-independent and the second one - $x_3$-independent. Thus, setting derivatives of the brackets equal zero we obtain the equations

$$\begin{cases} \dfrac{\partial \chi}{\partial x_1} = -\dfrac{\partial}{\partial x_1}\left(\dfrac{\theta^2}{\theta^2 - 1} H_1^2\right) \\[2ex] \dfrac{\partial \chi}{\partial x_3} = -\dfrac{\partial}{\partial x_3}\left(\dfrac{1}{\theta^2 - 1} H_1^2\right) \end{cases}$$

Since for any θ the potentiality condition

$$\frac{\partial}{\partial x_3}\frac{\partial \chi}{\partial x_1} - \frac{\partial}{\partial x_1}\frac{\partial \chi}{\partial x_3} = \frac{\partial^2}{\partial x_1 \partial x_3} H_1^2 = 0,$$

is satisfied due to Eq.(1.54),we can then reconstruct the function $\chi(x_1, x_3)$ up to some constant $\chi_0$, from its derivatives for any θ, in particular, for

$\theta = const\ (\theta > 1)$. Then, we reconstruct eikonal components B and S from the equations

$$\frac{\partial S}{\partial x_3} = \left( \frac{\theta^2}{\theta^2 - 1} H_1^2 + \chi \right)^{1/2}, \quad \frac{\partial B}{\partial x_1} = \left( \frac{H_1^2}{\theta^2 - 1} + \chi \right)^{1/2}$$

written down so that to obey complex-eikonal equation. Here an arbitrary constant $\chi_0$ of $\chi$ function is chosen so as to provide for the positive value of the expression in the brackets of previous equation.

In view of Eq.(1.53), transport equation (1.43) and field equation (1.44) admit separation of variables that yields us the equations

$$2B' \frac{\partial \ln A_0(x_1)}{\partial x_1} + B'' - C = 0 \tag{1.58}$$

$$2S' \frac{\partial \ln \gamma_0}{\partial x_3} + S'' + iC = 0 \tag{1.59}$$

$$\frac{1}{A_0} \frac{d^2 (A_0 F_p)}{dx_1^2} - 2kB' \frac{dF_p}{dx_1} + 2kn_0 \beta_p F_p = 0 \tag{1.60}$$

$$\frac{1}{\gamma_0} \frac{d^2}{dx_3^2} (\gamma_0 f_p) + 2ikS' \frac{df_p}{dx_3} - 2kn_0 \beta_p f_p = 0, \tag{1.61}$$

where the constant of separation for Eq.(1.44) is designated as $2kn_0\beta_0$. In this case the transport equations (1.58), (1.59) can be integrated even in a general form:

$$A_0(x_1) = \exp\left[ \frac{1}{2} \int_0^{x_1} \frac{C - B''}{B'} dx_1 \right] \tag{1.62}$$

$$\gamma_0(x_3) = \exp\left[ -\frac{1}{2} \int_0^{x_3} \frac{iC - S''}{S'} dx_3 \right] \tag{1.63}$$

Equation (1.60) is integrable only for specific types of the eikonal. In particular, equations (1.62), (1.63), (1.61), and (1.60) take the following form in a paraxial approximation

$$A_0(x_1) \cong const \tag{1.64}$$

$$B'' - C = 0, \quad B(x_1) = C(x_1 - x_{10})^2, \quad C = \frac{1}{2} B''(0) \tag{1.65}$$

$$f_p(x_3)=\exp\left[-i\beta_p n_o\int_0^{x_3}\frac{dx_3}{S'(x_3)}\right] \tag{1.66}$$

$$\frac{d^2F_p}{dx_1^2}-2kB'\cdot\frac{dF_p}{dx_1}+2kn_0\beta_pF_p=0 \tag{1.67}$$

It is from the known functions $B$, $S$, $A_0$, $f_p$, $F_p$ that one can find 1-D modes and eigenvalues, using the formulae of the kind (1.39)-(1.41):

$$\psi_0(x_1)=A_0(x_1)\exp[-kB(x_1)] \tag{1.68}$$

$$\psi_p(x_1)=\psi_0(x_1)F_p(x_1) \tag{1.69}$$

$$\gamma_p(x_3)=\gamma_0(x_3)f_p(x_3) \tag{1.70}$$

One-dimensional beams studied above are able to generate directly 2-D beams with the separable transverse coordinates. Under paraxial approximation it is convenient to consider separation of variables in the Cartesian coordinate system. Let us assume that we have already found 1-D function of the mode complex amplitude

$$W_p(x,z)=\exp[ikS(x,z)]\psi_p(x,z)\,\gamma_p(z) \tag{1.71}$$

that describe the mode complex amplitude, satisfy the Helmholts equation and is expressed in the Cartesian system instead of the curvilinear coordinates $x_1$, $x_3$. Similar functions $\tilde{W}_l(y,z)$ for y-direction can be found by the same procedure. The Helmholts equations, when taken in a paraxial approximation, change over to the parabolic equation, with separation of variables taken in the form :

$$W_{pl}(x,y,z)=W_p(x,z)\tilde{W}_l(y,z)\exp(-ikn_0z) \tag{1.72}$$

Therefore, all 1-D methods developed in this section turn out to be applicable to the 2-D modes. However, one should keep in mind that these modes have more complicated wavefront to be determined from the equation

$$S=S(x,z)+\tilde{S}(y,z)-n_0z=const, \tag{1.73}$$

where $S(x,z)$, $\tilde{S}(y,z)$ are the eikonals $S$ for the corresponding 1-D cases. The 2-D modes and their eigenvalues $\gamma_{pl}(z)$ are built on the basis of the 1-D ones, using the formulae

$$\psi_{pl}(x,y,z)=\psi_p(x,z)\psi_l(y,z) \tag{1.74}$$

$$\gamma_{pl}(z) = \gamma_p(z)\gamma_l(z) \tag{1.75}$$

that result from Eqs.(1.72), (1,73) and (1.35). Thus, the strategy of the transverse coordinates separation gives us a convenient method for constructing the modal beams of the various orders that are able to reconstruct themselves during propagation. It should be noted that the cross-structure of the mode, which is separable on $x,y$, can reproduce itself not only for the 2-D variant, Eq.(1.33), but for any 1-D beam-section produced by the plane parallel to the plane $(x,z)$ or $(y,z)$.

## 1.5. GAUSSIAN AND BESSEL MODES

### A. Gauss-Laguerre modes generalization

As [Solimeno,1986] have shown, an important kind of beams with rotational elliptical wavefronts in uniform medium exists and yields a degenerate elliptical coordinate system of an oblate rotational-symmetric ellipsoid:

$$\begin{cases} x = b\sin x_1 \,\mathrm{ch} x_3 \cos x_2 \\ y = b\sin x_1 \,\mathrm{ch} x_3 \sin x_2 \\ z = b\cos x_1 \,\mathrm{sh} x_3 \end{cases} \tag{1.76}$$

where

$$0 \le x_1 \le \frac{\pi}{2}, \quad 0 \le x_2 < 2\pi, \quad 0 \le x_3 < \infty.$$

Surfaces of the constant intensity fall are represented here by the surfaces of the coordinate $x_1 = const$, that is, by a one-sheet hyperboloid of revolution. According to [Korn, 1968] the Lame coefficients can be found from equalities

$$\begin{cases} H_1 = b\,(\mathrm{ch}^2 x_1 - \sin^2 x_3)^{1/2} \\ H_2 = b\sin x_1 \,\mathrm{ch} x_3 \\ H_3 = H_1 \end{cases} \tag{1.77}$$

[Solimeno , 1986] have described the S and B components of the eikonal corresponding to coordinates (1.76):

$$\begin{cases} S = n_0 b\,\mathrm{sh} x_3 \\ B = n_0 b\,(1 - \cos x_1) \end{cases} \tag{1.78}$$

Under a paraxial approximation, the solutions of Eqs.(1.47) through (1.50), (1.35) and (1.39) through (1.41) are found in analytical form and match to the well-known Gauss-Laguerre modes (1,20).

Gauss-Laguerre modes can be obtained from solving the general equations (1.49) and (1.50) taken in the coordinate system of Eq.(1.76).

Equation (1.76) written in the form

$$\begin{cases} r = b\sin x_1 \, \mathrm{ch} x_3 \\ z = b\cos x_1 \, \mathrm{sh} x_3 \end{cases}$$

with $r = \sqrt{x^2 + y^2}$ , allows us to draw a conclusion that the coordinate transformation occurs in the plane *r, z* and rotates around the *z*-axis. According to Eqs.(1.45) and (1.77), we have

$$\rho_1 = \rho_3 = b\sin x_1 \, \mathrm{ch} x_3, \quad \rho_2 = \frac{b(\mathrm{ch}^2 x_3 - \sin^2 x_1)}{\sin x_1 \, \mathrm{ch} x_3}.$$

Let us note that the functions $\rho_1$ and $\rho_3$ have the separable variables $x_1, x_3$ and the function $\rho_2$ is $x_2$- independent. The eikonal can be determined in an explicit form (1.78). We shall describe to the factor $F_{\mathrm{p}}$, the normalizing constant for the function $\psi_{\mathrm{p}}$ while choosing it arbitrarily for the remaining factors. In the transport equation (1.43), the variables $(x_1, x_3)$ can be separated with some $C$ constant and thus give a pair of equations for $A_0$ and $\gamma_0$. One can easily see that the only value, *C=1*, satisfies the requirement on $A_0$ to be finite at $x_1 = 0$ point. The transport equation is taking the form

$$\frac{\partial \ln A_0}{\partial x_1} = \frac{C - \cos x_1}{\sin x_1} \equiv \frac{\sin \frac{x_1}{2}}{\cos \frac{x_1}{2}}$$

$$\frac{\partial \ln \gamma_0}{\partial x_3} = -\frac{iC + \mathrm{sh} x_3}{\mathrm{ch} x_3} = -\frac{i \, \mathrm{ch} x_3}{1 + i \, \mathrm{sh} x_3}$$

That is readily integrable in the elementary functions:

$$A_0(x_1) \equiv \frac{1}{\cos^2 \frac{x_1}{2}} \tag{1.79}$$

$$\gamma_0(x_3) = \frac{1}{1 + i \, \mathrm{sh} x_3} = \frac{1}{\mathrm{ch} x_3} \exp(i \, \mathrm{arctg} \, \mathrm{sh} x_3) \tag{1.80}$$

The $x_2$ coordinate can easily be separated from $(x_1, x_3)$ coordinates in Eq.(1.44) for $F_{pl}$, by dividing both sides by $\rho_2$. After this we can also separate $x_1$ and $x_3$ coordinates :

$$F_{pl}(x_1,x_2)=F_p(x_1)\cdot F_l(x_2) \tag{1.81}$$

and obtain, the oscillator equation for $F_l$, of $x_2$:

$$\frac{d^2F_l}{dx_2^2}+l^2F_l=0, \quad F_l(x_2)=\exp(\pm ilx_2), \tag{1.82}$$

where $l^2$ denotes the constant of separation. The requirement on the $F_l(x_2)$, function to have single value with respect to each $x_2$ value, yields the requirement on $l$ number to be integer, $l=1,2,3,\ldots$.

Separation of the variables $x_1$ and $x_3$ from the constant designated as $-2kbn_0(2p+l)$, where $p$ is some number, gives us then

$$\frac{1}{A_0\sin x_1}\cdot\frac{\partial}{\partial x_1}\left(\sin x_1\frac{\partial(A_0F_{pl})}{\partial x_1}\right)-2kbn_0\sin x_1\frac{\partial F_{pl}}{\partial x_1}+ \tag{1.83}$$

$$+\left[2kbn_0(2p+l)-\frac{l^2}{\sin^2 x_1}\right]F_{pl}=0$$

$$\frac{1}{\gamma_0\operatorname{ch}x_3}\cdot\frac{\partial}{\partial x_3}\left(\operatorname{ch}x_3\frac{\partial(\gamma_0 f_{pl})}{\partial x_3}\right)+2ikbn_0\operatorname{ch}x_3\frac{\partial f_{pl}}{\partial x_3}-$$

$$-\left[2kbn_0(2p+l)-\frac{l^2}{\operatorname{ch}x_3}\right]f_{pl}=0 \tag{1.84}$$

where $A_0$, $\gamma_0$ are specified by Eqs.(1.79) and (1.80).

Equations (1.79) and (1.80), and (1.82) through (1.84) define non-paraxial Gauss-Laguerre beams propagating in a uniform medium of are similar to the solutions that [Landesman, 1988]. Equations (1.47) through (1.50), when taken in a paraxial approximation, led to the common Gauss-Laguerre modes. Actually, for the paraxial approximation considered, the functions $B$ and $S$ in Eq.(1.78) and the relationships for the coordinate transformation can be expanded in powers of $x_1$. Thus we can write down the following relationships:

$$r\cong bx_1\operatorname{ch}x_3\cong bx_1\sqrt{1+\operatorname{sh}^2 x_3},$$

$$z\cong b\operatorname{sh}x_3\left(1-\frac{x_1^2}{2}\right)\cong b\operatorname{sh}x_3,$$

$$\operatorname{sh}x_3\cong\frac{z}{b}\left(1+\frac{x_1^2}{2}\right)\cong\frac{z}{b}, \quad x_1=\frac{r}{b\sqrt{1+\frac{z^2}{b^2}}}, \tag{1.85}$$

$$B \cong n_0 b \frac{x_1^2}{2} = \frac{n_0 r^2}{2b\left(1+\frac{z^2}{b^2}\right)} = \frac{r^2}{k\sigma^2} \tag{1.86}$$

$$S \cong n_0 z\left(1+\frac{r^2}{2b^2\left(1+\frac{z^2}{b^2}\right)}\right) = n_0 z + \frac{n_0 r^2}{2R}, \tag{1.87}$$

where

$$R = z\left(1+\frac{b^2}{z^2}\right), \quad \sigma^2 = \sigma_0^2\left(1+\frac{z^2}{b^2}\right), \quad \sigma_0^2 = \frac{2b}{kn_0} \tag{1.88}$$

Furthermore, Eqs.(1.79), (1.80), and (1.85) give us

$$A_0(x_1) \cong 1,\ \gamma_0(x_3) = \frac{\sigma_0}{\sigma}\exp[-i\eta_0(z)] \tag{1.89}$$

and Eq.(1.84), under the (1.48) condition and $l << b/\lambda$, inequality yields

$$f_{pl}(x_3) = \exp[-i(2p+l)\eta_0(z)], \tag{1.90}$$

where

$$\eta_0(z) = \mathrm{arctg}\frac{z}{b} \tag{1.91}$$

One should note that in the paraxial approximation the factors $\gamma_0$, $f_{pl}$ depending on $x_3$ become dependent only of z and independent of $x$. Equation (1.83) takes the form

$$\frac{1}{x_1}\cdot\frac{\partial}{\partial x_1}\left(x_1\frac{\partial F_{pl}}{\partial x_1}\right) - 2kbn_0 x_1\frac{\partial F_{pl}}{\partial x_1} + \left[2kbn_0(2p+l) - \frac{l^2}{x_1^2}\right] = 0 \tag{1.92}$$

Using the conventional substitution

$$F_{pl} = t^{l/2}\cdot L_p^l(t),\ t = kn_0 x_1^2 \tag{1.93}$$

Eq.(1.92) is reduced to the canonical equation for the Laguerre polynomials

$$t\frac{d^2 L_p^l}{dt^2} + (l+1-t)\frac{dL_p^l}{dt} + pL_p^l = 0$$

that have the finite solutions in the zero and infinity points for the integer values only of $p = 0,1,2,\ldots$. On substituting derived equations (1.86), (1.87), and (1.89) through (1.93) into Eqs.(1.35) and Eqs.(1.39) through (1.41), and with Eq.(1.85) taken into account, we obtain the well-known formulae for the complex amplitude of Gauss-Laguerre modes described by [Yariv, 1985]

$$W_{pl}(\mathbf{x},z) = \psi_{pl}(\mathbf{x})\gamma_{pl}(z)\exp\left[ikn_0\left(z+\frac{\mathbf{x}^2}{2R}\right)\right] \tag{1.94}$$

$$\gamma_{pl}(z) = \exp[-i(r_{pl}+1)\eta_0(z)] \tag{1.95}$$

$$\psi_{pl}(\mathbf{x}) = E_{pl}\left(\frac{|\mathbf{x}|\sqrt{2}}{\sigma}\right)^l L_p^l\left(\frac{2\mathbf{x}^2}{\sigma^2}\right)\exp\left(-\frac{\mathbf{x}^2}{\sigma^2}\right)\exp(\pm il\alpha) \tag{1.96}$$

where α is the polar angle of the $\mathbf{x}$ vector, $L_p^l$ is the generalized Laguerre polynomial with $p,l = 0,1,2,\ldots$ described by [Korn, 1968]. Here

$$r_{pl} = 2p+l,\ \ E_{pl} = \frac{2}{\sigma\sqrt{l!C_{p+l}^l 2\pi}} \tag{1.97}$$

is the normalizing constant.

In some cases, the modes of Eq.(1.96) are written in a real form including $\sin(l\alpha)$ and $\cos(l\alpha)$ instead of $\exp(\pm il\alpha)$.

The value z=0 corresponds to the waist of the beam, 2b is the confocal parameter of the Gaussian beam, and the parameters σ, R and η are specified by Eqs.(1.88) and (1.91). Thus it is, for the introduced elliptical coordinate system of that the complex eikonal method in a paraxial approximation yields the conventional Gauss-Laguerre modes.

On the basis of the quasi-classical approximation described by [Landau, 1974], one can easily evaluate the diameter $2a$ $2\sigma\sqrt{(2p+l)_{\max}+1}$ for the Gauss-Laguerre modes numbered so that $2p+l \le (2p+l)_{\max}$ in the case of large $p$- and $l$- numbers.

**B. Generalized Gauss-Hermite modes**

For $w_0 = const$, one of the solutions of Eq.(1.57) is given by

$$X(U) = b\sin U \tag{1.98}$$

On changing over to the real and imaginary parts, we get

$$\begin{cases} x = b\sin x_1 \operatorname{ch} x_3 \\ z = b\cos x_1 \operatorname{sh} x_3 \end{cases}$$

$$H_1^2 = b^2(\mathrm{ch}^2 x_3 - \sin^2 x_1) = H_3^2$$

Here, the eikonal has the same form, of Eq.(1.78), as in the case of Gauss-Laguerre mode. As shown below, paraxial equations together with Eq.(1.72) give the solution in the form [Yariv, 1985] of the Gauss-Hermite modes. For a non-paraxial variant, Eq.(1.98) will give the 1-D Gauss-Hermite mode's generalization. Other generalizations of the 1-D modes may be based on the conformal coordinate transformations different from those described by Eq.(1.98). By virtue of Eq.(1.57), such transformations are generated by the real-valued functions $\omega_0^2$ varying with respect to $x_1$ and $x_2$.

The 1-D Gauss-Hermite modes can be deduced from Eqs.(1.68) through (1,72), and (1.74) that use the functions $S$, $B$, $A_0(x_1)$, $\gamma_0(x_3)$, $f_p(x_3)$, $F_p(x_1)$.

According to equations (1.62), (1.63), and (1.60),(1.61) we have

$$A_0(x_1) = \frac{1}{\cos\frac{x_1}{2}} \tag{1.99}$$

$$\gamma_0(x_3) = \frac{1}{\sqrt{1+i\,\mathrm{sh}\,x_3}} = \frac{1}{\sqrt{\mathrm{ch}\,x_3}}\exp\left(-\frac{1}{2}\,i\,\mathrm{arctg}\,\mathrm{sh}\,x_3\right) \tag{1.100}$$

$$f_p(x_3) = \exp(-ip\,\mathrm{arctg}\,\mathrm{sh}\,x_3), \tag{1.101}$$

where

$$p = \frac{\beta l}{b} \tag{1.102}$$

The $F_p$ function is determined from the ordinary differential equation of the second order with varying coefficients, see Eq.(1.60):

$$\cos\frac{x_1}{2}\cdot\frac{d^2}{dx_1^2}\left(\frac{F_p}{\cos\frac{x_1}{2}}\right) - 2kbn_0\sin x_1\frac{dF_p}{dx_1} + 2kn_0 b_p F_p = 0 \tag{1.103}$$

For the paraxial approximation we can write

$$\frac{|x|}{z} << 1,\quad \cos\frac{x_1}{2} \cong 1;\quad \sin x_1 \cong x_1,\quad A_0 \cong 1. \tag{1.104}$$

In this case Eq.(1.103) for $F_p$ (on substitution the variables as $t = \sqrt{kbn_0}\,x_1$) is reduced to the standard equation for the Hermite polynomials, described by [Korn, 1968]

$$\frac{d^2F_p}{dt^2}-2t\frac{dF_p}{dt}=-2pF_p,$$

Last equation has the finite for $t=0$ and $t\to\infty$ solutions only for the integer values of $p=0,1,2,\ldots$. The solutions for $F_p$ thus can be written in the form:

$$F_p = const\ H_p\left(\sqrt{bkn_0}x_1\right) \tag{1.105}$$

Moving further, using Eqs.(1.99) through (1.103) and condition (1.104), we obtain Eqs.(1.85) through (1.88), where $\sigma^2$ and R are specified by Eqs.(1.88) and (1.91), as is the case for the Gauss-Laguerre modes. For $\gamma_0$ and $f_p$, we have

$$\gamma_0(x_3)=\sqrt{\frac{\sigma_0}{\sigma}}\exp\left(-i\frac{1}{2}\mathrm{arctg}\frac{z}{b}\right) \tag{1.106}$$

$$f_p(x_3)=\exp\left(-ip\,\mathrm{arctg}\frac{z}{b}\right). \tag{1.107}$$

Let us note that in a paraxial approximation the values $\gamma_0$ and $f_p$ depend only on z not on x.

Finally, the 2-D modes and eigenvalues are equal to the product of the corresponding 1-D modes and, by virtue of Eqs.(1.105) - (1.106), can be written in the form (1.94), (1.95) with [Yariv, 1985]

$$\psi_{pl}(\mathbf{x})=E_{pl}H_p\left(\frac{\sqrt{2}x}{\sigma}\right)H_l\left(\frac{\sqrt{2}y}{\sigma}\right)\exp\left(-\frac{\mathbf{x}^2}{\sigma^2}\right), \tag{1.108}$$

$$r_{pl}=p+l \tag{1.109}$$

and

$$E_{pl}=\frac{1}{\sigma}\sqrt{\frac{2}{\pi 2^{p+l}p!l!}} \tag{1.110}$$

where $H_p(\cdot)$ is the Hermite polynomial of the $p$-th order.

With the help of a quasi-classical approximation [Landau, 1974] for large numbers of p and l, it is easy to estimate the size $2a\times 2b$ of the region comprising modes up to the order of $(p,l)$:

$$2a\ \ 2b\sqrt{p_{\max}+1/2},\ \ 2b\ \ 2\sigma\sqrt{l_{\max}+1/2}$$

where $p_{\max}$ and $l_{\max}$ are the maximum values of the numbers $p$ or $l$, respectively, for the group of modes under consideration.

**C. Gaussian modes in passive cavities and graded-index fibers**

Gaussian modes include the Gauss-Hermite and Gauss-Laguerre modes with several common properties. Strictly speaking, the Gaussian modes are orthogonal on the infinite domain. However, because of the quick amplitude fading, they can be considered to be orthogonal on some finite domain G. The Gaussian modes describe modes, with not very high orders that occur in the open passive cavity with spherical mirrors. Furthermore, Gaussian modes characterize the modes of graded-index optical fiber with the refractive index' quadratic profile as well as others lens-like waveguide media.

Gauss-Laguerre modes match to annular mirrors or graded-index quadratic optical fiber with the round cross-section. The Gauss-Hermite modes match to rectangular mirrors or to the optical fiber with the rectangular cross-section. Paraxial approximation condition needs to be satisfied for using the Gaussian modes in the cavity:

$$\frac{2a2b}{R_1R_2} << 1 \quad \text{or} \quad \frac{(2a)^2}{R_1R_2} << 1$$

were $2a \times 2b$ are the aperture sizes of the largest spherical mirror, $R_1, R_2$- radii of curvature of spherical mirrors

For a cavity, the parameter $\sigma_0$ can be found from the equation described in the handbook [ Pressley, 1971]:

$$\sigma_0 = \left(\frac{\lambda}{\pi}\right)\frac{\left[l(R_1 - l)(R_2 - l)(R_1 + R_2 - l)\right]^{1/2}}{(R_1 + R_2 - 2l)}$$

and the waist itself is disposed in a cavity at a distance

$$z_2 = \frac{l(R_1 - l)}{R_1 + R_2 - 2l}$$

from the output mirror.

For the propagation in a fiber with the quadratic refractive index described in the form

$$n^2(r) = n_1^2\left(1 - 2\Delta\frac{r^2}{a^2}\right),$$

we have

$$\gamma_{pl}(z) = \exp(i\beta_{pl}\, z),$$

and the value of $\beta_{pl}$ and $\sigma_0$ are found from the formula given by [Yariv, 1985]

$$\beta_{pl} = \left[k^2 n_0^2 - \frac{4}{\sigma_0^2}(r_{pl} + 1)\right]^{1/2}$$

$$\sigma_0 = \left(\frac{\lambda a}{\pi n_1}\right)^{1/2} \left(\frac{1}{2\Delta}\right)^{1/4} = const; \quad R = \infty$$

The Gaussian beams manifest a remarkable property of conservation of their type, while propagating not only in free-space but through lenses, too. According to [Yariv, 1985]Gaussian modes propagating through a optical Fourier-stage with the focal length $f_0$, change the parameters σ, R=∞ to $\sigma_F$ and $R_F$ where

$$\sigma_F = \frac{\lambda f_0}{\pi \sigma}, \quad R_F = \infty$$

The phase of the beam changes in this case by the following value:

$$kf_0 + \arg(i^{2p+l}) = kf_0 + r_{pl}\frac{\pi}{2}$$

Thanks to the Gaussian mode invariance under propagation within the Fourier-stage, the former can be used as an optical projection system that transfers a magnified multimodal beam from the fiber output to the modan or, vice versa, transfers an attenuated and focused beam from the modan to the fiber.

**D. Bessel modes**

According to [Unger, 1977], Bessel modes are described by equations

$$\psi_{pl}(\mathbf{x}) = E_{pl} J_l\left(\rho_{pl}\frac{|\mathbf{x}|}{a}\right)\exp(\pm il\alpha), \quad |\mathbf{x}| \le a; \quad p,l = 0,1,2,...$$

$$E_{pl} = \frac{1}{a\sqrt{\pi} J_l'(\rho_{pl})},$$

where $J_l(\bullet)$ is the Bessel function of integer order, $\rho_{pl}$ is the $p$-th zero of the function $J_l(\bullet)$.

**E. Sinusoidal modes**

Sinusoidal modes are trigonometric base-functions that are orthogonal on the rectangular domain G:

$$\psi_{pl}(\mathbf{x}) = \frac{1}{\sqrt{G}}\exp(i2\pi\mathbf{v}^{(pl)}\mathbf{x}), \quad \mathbf{x} \in G, \quad p,l = 0,\pm 1,...$$

$$\mathbf{v}^{(pl)} = (v_x^{(p)}, v_y^{(l)}), \quad v_x^{(p)} = pv_x^{(1)}; \quad v_y^{(l)} = pv_y^{(1)}; \quad v_x^{(1)} = \frac{1}{2a}; \quad v_y^{(1)} = \frac{1}{2b}$$

$$G = \{\mathbf{x}: |x| \le a,\ |y| \le b\}, \quad |G| = 2a2b,$$

Such functions describe, for example, the modes with not very high orders that exist in a plane-parallel cavity with rectangular mirrors of the size $2a \times 2b$.

# Chapter 2

# MODANS AS THE COMPUTER GENERATED HOLOGRAMS

## 2.1. PHYSICAL OBSERVABILITY OF MODES AND THE CONCEPT OF MODANS.

Coherent field's modal structure is characteristic for a variety of waveguide media such as passive and active laser cavities, lens-like media, optical fibers. All the experience gained in optics and quantum electronics shows that only a principal mode and some or lower-order modes could be excited as single ones. Other modes really can exist only in "mixtures" or modal groups with various modal power distributions. Let us discuss an important optical problem: whether the modes are real physical objects and some kind of optical standards or they are only the mathematical orthogonal base functions for representing multimodal beams of coherent light. At one time, the same problem was under discussion with reference to the temporal color spectrum: are there physically existent monochromatic primary light harmonics or they are simply convenient mathematical expansions of light into a series of sinusoidal harmonics. From a modern standpoint, the following answer proves to be justified: monochromatic harmonics do exist because:

a) they propagate in free space without changing their structure or wavelength;

b) there are spectral devices enabling of the selection, excitation, observation, and measurement of harmonics in the form of spectrum to be done.

The property (b) of monochromatic light components is realized by using spectrographic instrumentation for the spectral analysis of light. Monochromators, light filters, and laser sources implement selective excitation of the single monochromatic components or their groups.

An operation of conventional spectral devices can be interpreted as the decomposition of light beams into plane or spherical beams of different tilts and wavelengths (on passing a grating or a prism) (see Fig. 2.1). Those waves can be treated as free space modes numbered using the parameters of the tilt (equivalent to the transverse modal indices) and the frequency of radiation (equivalent to a longitudinal modal index). Plane and spherical monochromatic waves also satisfy the property (a), namely, in the course of their propagation in free space they do not blur in the color spectrum, preserve their individual spatial form and obtain only phase delays. Thus, spectral harmonics possessing the (a) and (b) properties exist as physical objects and represent the optical standards of color spectrum.

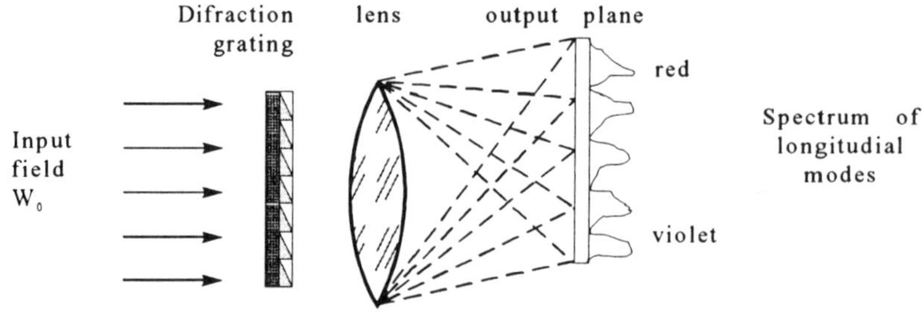


**Figure 2.1.** Diffraction grating that splits the beam into the temporal spectrum of colors ("longitudinal" modes).

One can do an out-of-cavity selection for some modes of coherent light with different wavelengths by using high-resolution conventional spectral devices with a highly dispersing optical element in the form of a prism or an echelette. However, various fiber modes are degenerated with respect to the wavelength. Nevertheless they have differences in the cross-section spatial structure of different modal beams.

The modes of coherent light can be treated as "generalized spectral" components since they propagate in their medium with their own modal number and individual form and acquire only the phase delay, see Eqs.(1.1) and (1.16) as for the (a) property of the temporal spectrum. Thus, the solution of the problem of modes' physical existence depends only on the possibility to develop the devices capable of forming the standard modal beams as well as the devices for measuring the modal composition of available multimodal beams (the property (b)). Such devices will be able to perform an individual excitation, measurement and selection of the modes of coherent light. Conventional optical elements turn out to be unable to form standards for the arbitrary selectively excited modal groups of coherent light. One of the approaches that might yield the solution of the modes' selective formation is development of a reference laser for generating the required modes. However, in this case, significant difficulties are met with in providing the proper form of the field distribution in a laser beam. For example, modes with the circular symmetry (the Gauss-Laguerre modes, in particular) can arise only in the case of strictly axial symmetry of the cavity. Also, it is of considerable difficulty to excite just the required modes, especially of higher orders.

As [Lissak, 1990] and [Pressley, 1971] have described, the Gaussian principal mode is readily selected by the simple diaphragm if some power losses are tolerable. The Littrow gratings are utilized for the inside-cavity modal selection [Solimeno, 1986]. However, their disadvantage is that they are suitable only for selecting either principal or "longitudinal" modes that, in effect, are not the modes but only the monochromatic components of the conventional chromatic spectral dispersion.

For the purpose of selecting the lower-order modes, [Berdague, 1982] proposed employing such complex diaphragms with the gas cells that reproduce the form of the amplitude zeros of modes and are not the standards of separate modes since they reconstruct many parasitic modes. Mode excitation in optical fibers can be formed to a certain extent by the systems of prism-aided launching devices [Stewart, 1975], and devices utilizing diffraction gratings and reflectors [Nishi, 1987]. Fibers' connections and microbendings that have been discussed by [Gowar, 1984; Grau, 1980; de Fornel, 1983; Krivoshlykov, 1983][Krivoshlykov, 1987] also affect the modal power distribution. The mentioned approaches feature the limited selectivity and they are capable of exciting the groups of modes rather than separate modes.

It seems essential to note that we are able to set up a correspondence between a conventional optical component - prism, and a diffractive optical component, an echelette grating that, in fact, is the kinoform of a prism [Lesem, 1969] and has similar or even better operational capabilities (see Fig. 2.1). Nevertheless, there is not any known available conventional optical element purposed to select modes in cavities and optical fibers, that is, to resolve the problem of spectral analysis in terms of modes. There is not also any available conventional optical element capable of forming the standards of separate modes as well as the required modal groups of coherent light.

Thus, it appears that a problem of the mode selective analysis remains unsolved in terms of conventional optical systems since there are no optical elements or spatial filters that are matched to the modes of lasers and optical fibers.

In view of all said above, the development of modans - novel diffractive optical elements selecting the modes of coherent light - appears to be a matter of current interest (see Fig. 2.2). Modans are to comprise the standards of modes.

Further studies may yield modans created on the basis of the purely physical approaches. However, as we see it, nowadays the modans, which are generated with the aid computers, prove to be most realistic.

Among prerequisites for the use of computer generated artificial mode standards are also digital methods for investigating the complex amplitude or intensity of multimodal beam in the near-field or far-field zone [Diado, 1979; Kitayama, 1979; Shigezawa, 1978]. These methods give the possibility to restore the modal power distribution by computerized image processing. However, they require high-quality scanning, analog-to-digital conversion as well as large-volume digital processing of diffraction images and interferograms. It must be emphasized that for the above mentioned digital methods the mode standard is actually specified within the computer. It seems natural to convert a digital standard to the physical object in the form of the optical element - modan.

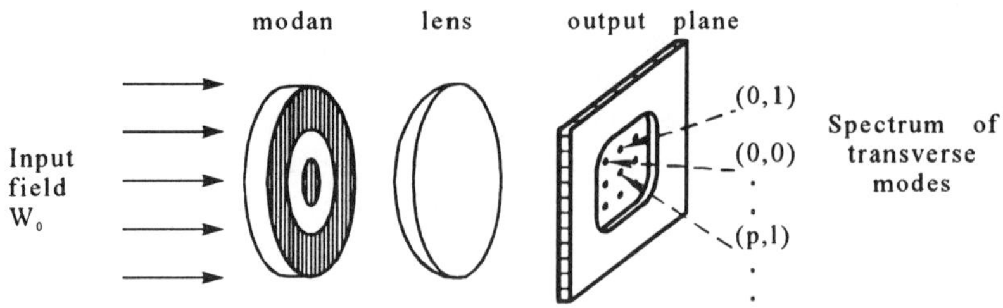


**Figure 2.2.** The modan forming the "generalized" spectrum of "transverse" modes of coherent light .

Some ideas for mode selection can be derived from the optical systems calculating coefficients of expansions of the 2-D light fields [Gibin, 1972; Lee, 1974]. Thus, modans offer direct interconnection between the mathematical mode concept and real optical systems via computer.

The first computer generated optical elements aimed at analyzing and forming the transverse-mode light composition, were developed by the present authors in 1981-1982 [Golub, 1982] in the form of the amplitude spatial filters and holographic filters. They have successfully passed the first testing by in 1983, 1984[Golub, 1983][Golub, 1984] and more detailed experimental investigation by G.V.Uvarov and S.V.Karpeev later [Garitchev, 1985; Golub, 1988], [Golub, 1989b; Garitchev, 1990] [Golub, 1991b]. In 1983 the works of other researchers [Bartelt, 1983] dealing with this problem were published, where the well-known in digital holography Lohmann's method of binary cells is applied to generating the modal spatial filters.

Recently, we have developed the phase modans [Golub, 1988] that possess enhanced energy efficiency and are capable of multibeam operating that makes it possible to form several modal beams simultaneously.

We propose the methods of diffractive optics and digital holography as the best approach to develop a novel computer generated diffractive modan and the mode standards. Digital holography methods involve following steps. Modans are first computed by solving eigenvalue equation (1.1), by introducing the models of the required modal beam and by coding the complex transfer function into the phase one. The modans' masks are then generated by scanning and converted to the phase-recording medium in the form of the reflective or refractive diffraction microrelief.

In fact, modans are diffractive optical elements similar to diffraction gratings but allowing of separation of radiation into "transverse" spatial modes of waveguide medium rather than in longitudinal modes. Although we can put the conventional diffraction grating in correspondence with the conventional prism, it seems to be impossible to associate any known optical element with the modan. Modans offer an opportunity to develop perfectly novel optical devices to analyze and form any needed modal power distribution of light beams.

There is a special case, where we analyze random optical fields (for instance, those reflected from a rough surface, scattered by a spatial micro particle ensemble or transmitted through the random medium). It is the matter of interest to separate their statistically independent components, which are similar in some sense to the conventional modes. Special expansions of random fields in terms of eigenfunctions of their correlation operator [Karhunen, 1947; Klovskiy, 1976] are well known in mathematical statistics and in theory of control. They provide uncorrelated coefficients of expansion and can be treated as some analogs of modal expansions (see Appendix 4).

In view of further consideration, it appears convenient to introduce a formally generalized concept of the "modan." For this purpose, consider the input $(\mathbf{u}, z_0)$ and output $(\mathbf{x}, z_0 + F_0)$ planes of some optical device (see Fig. 2.3). Suppose, there is an incident field in the waveguide medium with the complex amplitude $w(\mathbf{u}, z_0)$ that enters the input plane of an optical device. The device performs selection of the "transverse" modes corresponding to the waveguide medium so that the weight coefficients of the modes become observable separately from one another, and each coefficient is observed in that point of the output plane $\mathbf{x}$=(x,y), which is known in advance. Let us call such a device a "modan" (the modal analyzer).

Let us consider the situation to be opposite to some extent to that discussed above. Assume that the input plane $\mathbf{u}$ of an optical device is illuminated with the plane wave. The device forms in the output plane $\mathbf{x}$ the field $w(\mathbf{x}, F_0)$ consisting of a predetermined set of transverse modes. Such a device should be called a mode formation. As shown below, the mode formator is realized on the basis of modan. The problem of the modans' physical and technical realization is dealt with below in this work.

## 2.2. MULTIFOCAL AND MULTIBEAM HOLOGRAPHIC OPTICAL ELEMENTS

Multifocal optical components could have interesting applications in the field of optical information processing and image formation. Multifocal HOEs are also similar to the holograms with a number of exposures or to the holograms of a set of objects. They are indeed matched spatial filter for several objects in parallel. In this book we discuss such elements in view of their use in the parallel simultaneous operation, with several modes involved. Holographic optical elements (HOE) offer new possibilities in this area.

Each partial beam describing only one object is superimposed on a spherical or another focusing wave with a tilt (see Fig. 2.4). A multifocal optical element can be got on the basis of coupling a conventional lens with a "multibeam" HOE that reconstructs a set of quasi-plane waves upon illuminating with a single plane wave (see Fig. 2.5).

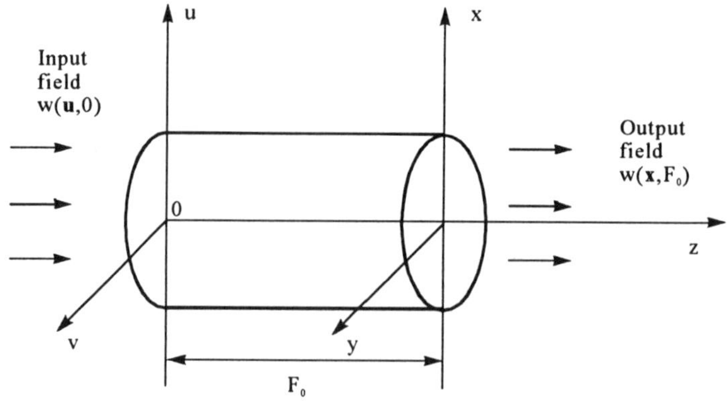


**Figure 2.3.** Optical scheme for the modan's operation.

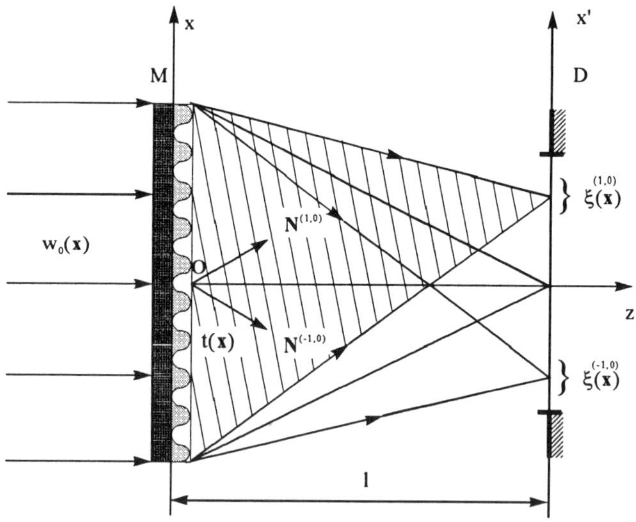


**Figure 2.4.** Multifocal HOE separating the beam in terms of quasi-spherical waves. Lens is "superimposed" in this element. $M$ - modan with $\Gamma$ transfer function; $w_0(\mathbf{x})$-illuminating beam; $N^{(1,0)}$and $N^{(-1,0)}$-unit vectors showing the directions of (1,0) and (-1,0) diffraction orders; $\xi^{(1,0)}(\mathbf{x})$ and $\xi^{(-1,0)}(\mathbf{x})$-partial light beams reconstructed from modan; $D$-diaphragm, taking the needed diffraction orders only; $l$ - focal distance.

Integrated operation of HOE-and-lens combination can take place in a Fourier-stage (see Fig. 2.6).After this we can treat modan depicted in Fig 2.4. as containing some lens in it implicitly.

Now we shall consider only multibeam HOE that works in terms of plane-wave generation in such a multibeam HOE, that each object wave is superimposed on a plane wave and then recorded in the form of the plane-wave phase modulation.

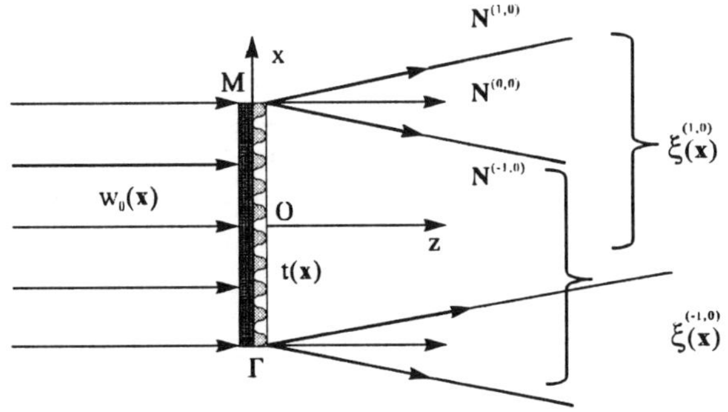


**Figure 2.5.** Multibeam HOE separating the beam in terms of quasi-plane waves. $M$ - modan with Γ transfer function; $w_0(\mathbf{x})$-illuminating beam; $N^{(1,0)}$ and $N^{(-1,0)}$ - unit vectors showing the directions of (1,0) and (-1,0) diffraction orders; $\xi^{(1,0)}(\mathbf{x})$ and $\xi^{(-1,0)}(\mathbf{x})$-partial light beams reconstructed from modan.

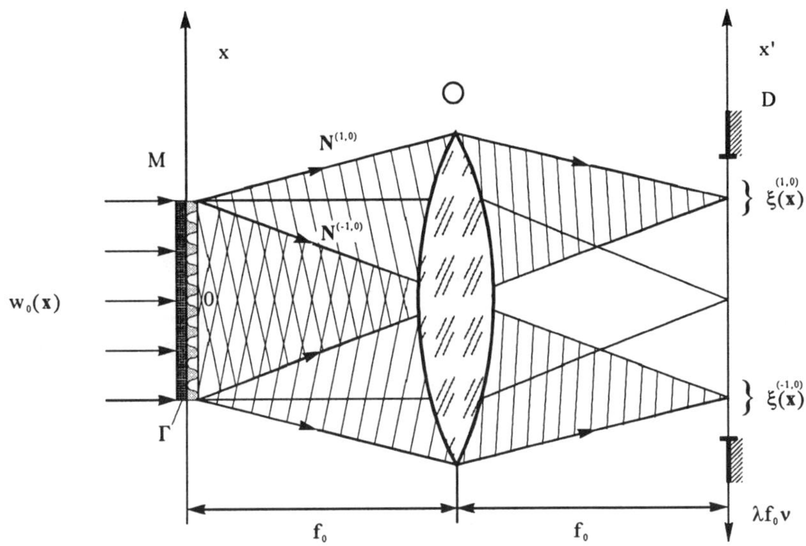


**Figure 2.6.** Multibeam optical element integrated into a Fourier-stage in order to reconstruct images of modes. $M$ - modan with Γ transfer function; $w_0(\mathbf{x})$-illuminating beam; $N^{(1,0)}$and $N^{(-1,0)}$-unit vectors showing the directions of (1,0) and (-1,0) diffraction orders; $\xi^{(1,0)}(\mathbf{x})$ and $\xi^{(-1,0)}(\mathbf{x})$- partial light beams reconstructed from modan; $O$ - objective; $D$ - diaphragm, taking the needed diffraction orders only, $f_0$ -focal length.

Such HOE represents a mathematical superposition of several object beams with corresponding spatial carrier frequencies and, in fact, comprises a set of diffraction gratings. The vectors of the carrier spatial frequencies are chosen as differently oriented in order to separate the branches.

Multibeam HOEs can be generated from singlebeam ones by means of superimposing the branches using optical holography and interferometry techniques. Let us call such HOEs the optical-digital or hybrid HOEs. For example, aiming to make a thin hologram that includes several differently oriented diffraction gratings, one should generate a set of HOEs in the form of separate object holograms characterized by the carrier spatial frequency that is small enough (10 lines/mm, [Bartelt, 1982] and [Wang, 1990]. [Denisyuk, 1986] have made a thick multibeam HOE in the form of the 3-D hologram with high diffraction efficiency. L.P.Yaroslavskii and N.S.Merzlyakov [Yaroslavskii, 1980] have reviewed other approaches to the problem of fabrication of optical-digital holograms. The use of optical-digital holograms reduces the requirements for resolution of masks. However, the procedure of optically generating such a multibeam HOE proves to be complicated and multistage. It seems expedient that the stages of repeated exposure should be directly introduced into the algorithm of computer generation. Computer generated optical elements (CGOEs) are of particular interest in this section as the best way to realize multibeam HOEs. The multifacet splitting of a beam in a way suggested in [Doskolovich, 1991] by dividing a synthesized optical element into facets does not always accord to the objects' configuration and is sensitive to the degree of the beam uniformity. [Gamble, 1990] have proposed a different approach to the problem that lies within the frames of the image multiplication. Many diffractive approaches for beam multiplexing have been developed recently in the scientific area of optical interconnections [Lee, 1993; Morrison, 1993; Vasara,1993; West, 1993; Zhang, 1993]. Binary zone plates (see [Papoulis, 1968]) are only suitable for a narrow class of problems of multibeam image formation, mainly for the similar image in all branches. The approach of phase nonlinearity in multibeam operation has been investigated in [Golub, 1992b].

Here we want to have several partial beams originating from the same aperture with different angles. It is important to achieve different partial images reconstructed from the partial beams. Thus in this section we discuss an angular separation of partial beams taking place in multibeam HOEs. Let us calculate the complex transfer function for a multibeam HOE forming $L$ partial beams enumerated with the double index $\boldsymbol{k}$ that runs the set $K$ comprising $L=|K|$ pairs of indices (here, $|K|$ denotes the number of elements in a finite set $K$). A relative direction of propagation for each partial beam is specified with the aid of the 2-D vector of the spatial carrier frequency $\mathbf{v}^{(\mathbf{k})}, \mathbf{k} \in K$. The portion $\kappa^{(\mathbf{k})}$ of the illuminating light flux that should be guided into each partial beam number $\mathbf{k} \in K$ as well as the interbranch phase shifts $\alpha^{(\mathbf{k})}$ are specified. The condition must be satisfied

$$\sum_{\mathbf{k} \in K} \kappa^{(\mathbf{k})} = 1 \tag{2.1}$$

To provide convenient form in representation of the equation, let us introduce the designations for the scalar product and norm

$$(w_1, w_2) = \int_G w_1(\mathbf{x}) w_2^*(\mathbf{x}) \rho_0(\mathbf{x}) d^2\mathbf{x}, \qquad \|w\| = [(w,w)]^{1/2}$$

in the space of quadratic integrable functions [Rudin, 1973], where $d^2\mathbf{x} = dxdy$, $*$ is the symbol of the complex conjugation, and $\mathbf{x}(x,y)$ are the coordinates of the wavefront point projection onto the plane perpendicular to the direction of mode propagation, $\rho_0(\mathbf{x})$ is the weight-function.

The functions $w_1$ and $w_2$ have the meaning of the complex amplitude values of wavefront. However, in the course of computation of the scalar products and norms of the modes with a common wavefront, Eq.(1.35), the factor $\exp(ikS)$ is canceled with its conjugate value. Hence, under $w_1$ and $w_2$ one can imply the complex amplitude values taken directly in the plane $(x,y)$. The square of the norm

$$\|w\|^2 = \int_G |w(\mathbf{x},z)|^2 \rho_0(\mathbf{x}) d^2\mathbf{x}$$

is proportional to the power of the light passing through the wavefront within the limits of the domain $G$. Below we shall also use the designations $(\cdot,\cdot)$ and $\|\cdot\|$ for the Cartesian scalar product and the norm of finite-dimensional vectors, respectively. Let us consider the holographic element to be a thin optical element with the 2-D Cartesian coordinates in its plane (see Fig. 2.5). The $z$-axis is drawn from the HOE center in parallel to the direction of propagation. The transforms are performed in the domain $\mathbf{x} \in G$. An illuminating beam forms the complex amplitude $w_0(\mathbf{x})$ immediately behind the HOE plane. The aim of the multibeam HOE is to create, just behind its plane, such complex amplitude $t(\mathbf{x})$ that describes $L$ partial light beams $\xi^{(\mathbf{k})}(\mathbf{x})$ superimposed on the carrier frequencies $\mathbf{v}^{(\mathbf{k})}$ with the given amplitude $\sqrt{\kappa^{(\mathbf{k})}}$ and initial phase $\alpha^{(\mathbf{k})}$. A total output beam just behind the HOE plane can be specified as a superposition of several beams with varying tilt according to the equation

$$t(\mathbf{x}) = \sum_{\mathbf{k} \in K} \mu^{(\mathbf{k})} \frac{\xi^{(\mathbf{k})}(\mathbf{x})}{\left\|\xi^{(\mathbf{k})}\right\|} \exp[i2\pi \mathbf{v}^{(\mathbf{k})} \mathbf{x}] \tag{2.2}$$

that generalizes the equations proposed by [Lee, 1974]. Here the following designation is used

$$\mu^{(\mathbf{k})} = \sqrt{\kappa^{(\mathbf{k})}} \exp[i\alpha^{(\mathbf{k})}]. \tag{2.3}$$

The beam $t(\mathbf{x})$, Eq.(2.2), with fully separated branches represents a superposition of $L$ partial beam propagating in various directions characterized by the unit vectors

$$\mathbf{N}^{(\mathbf{k})} = \left( \lambda \mathbf{v}^{(\mathbf{k})}, \sqrt{1-\lambda^2 (\mathbf{v}^{(\mathbf{k})})^2} \right), \mathbf{k} \in K.$$

The value of $\|t\|^2$ is equivalent to the total power of $L$ branches of the beam **t** and can be represented in the form.

$$\|t\|^2 = \sum_{\mathbf{k} \in K} \kappa^{(\mathbf{k})} + \sum_{\mathbf{k} \in K} \sqrt{\kappa^{(\mathbf{k})} \kappa^{(\mathbf{k}')}} \exp[i\alpha^{(\mathbf{k})} - i\alpha^{(\mathbf{k}')}] \times$$
$$\times \frac{1}{\left\|\xi^{(\mathbf{k})}\right\| \cdot \left\|\xi^{(\mathbf{k}')}\right\|} \int_G \xi^{(\mathbf{k})}(\mathbf{x}) \xi^{(\mathbf{k}')*}(\mathbf{x}) \exp[i 2\pi (\mathbf{v}^{(\mathbf{k})} - \mathbf{v}^{(\mathbf{k}')}) \mathbf{x}]\, d^2\mathbf{x}$$

The branches will not overlap subject to the condition that "interbranch" integrals are small. For this purpose the period $\left|\mathbf{v}^{(\mathbf{k})} - \mathbf{v}^{(\mathbf{k}')}\right|^{-1}$, $\mathbf{k}' \neq \mathbf{k}$ the oscillating factor of is to be small compared with the minimum distance $\Delta l$ between zeros of the function $\xi^{(\mathbf{k})}(\mathbf{x}) \cdot \xi^{(\mathbf{k}')*}(\mathbf{x})$, $\mathbf{k} \in K$, $\mathbf{k}' \in K$. Thus, the beam separation condition takes the form

$$\min_{\mathbf{k} \neq \mathbf{k}'} \left|\mathbf{v}^{(\mathbf{k})} - \mathbf{v}^{(\mathbf{k}')}\right| >> \frac{1}{\Delta l} \qquad (\mathbf{k} \in K, \mathbf{k}' \in K) \qquad (2.4)$$

Therefore, the frequencies $\mathbf{v}^{(\mathbf{k})}$ of branches are to be different enough so that the magnitude $\left|\mathbf{v}^{(\mathbf{k})} - \mathbf{v}^{(\mathbf{k}')}\right|$, for any $\mathbf{k}' \neq \mathbf{k}$, is much more as compared with the maximum spatial frequency of the objects under studies. In that case the diffraction orders that are superimposed in Eq.(2.2), prove to be separated in angles and the total beam power can be written in the form

$$\|t\|^2 = \sum_{\mathbf{k} \in K} \kappa^{(\mathbf{k})} = 1.$$

## 2.3. MODANS ON CROSSED DIFFRACTION GRATINGS

The modans' operation in several branches requires the formation of several modes or modal groups simultaneously. A successive replacement in an optical setup of the singlebeam modans comprising the standard of a single mode or a single modal group is the simplest but rather inconvenient method. A promising approach is using a single modan forming several beams each of them corresponds to one branch of the mode analysis. The said above necessitates the development of the multibeam modan representing an optical element, which splits illuminating light beam into $L$ branches and allows parallel and selective excitation and analysis of several modes or modal groups to be done. Multibeam modans might be helpful in the tasks where branching or, vice versa, mixing of light beams is needed.

In terms of the communication theory, a partial beam is one of the channels for data communication. A multibeam modan implements spatial-frequency multiplexing of branches as channels and, thus, can be referred to as a "multibeam" modan also. Let us formulate a general problem of synthesis of the multibeam modan as that of synthesis of the multibeam hologram discussed in section 2.2.

A required modal composition of each partial beam numbered **k** is given by the set $\Xi^{(\mathbf{k})}$ of the coefficients $\xi_{\mathbf{p}}^{(\mathbf{k})}$ of the modal function $\psi_{\mathbf{p}}$ in the beam of $k$-th branch

$$\Xi^{(\mathbf{k})} = (\xi_{\mathbf{p}}^{(\mathbf{k})}, \mathbf{p} \in L^{(\mathbf{k})}), \quad \mathbf{k} \in K$$

The coefficients $\Xi^{(\mathbf{k})}$ contain the power distribution between modes of one branch and intermodal phase shifts; characterized by the following values:

$$\frac{\left|\xi_{\mathbf{p}}^{(\mathbf{k})}\right|^2}{\left|\Xi^{(\mathbf{k})}\right|^2}; \qquad \arg\xi_{\mathbf{p}}^{(\mathbf{k})}, \quad \mathbf{p} \in L^{(\mathbf{k})}, \quad \mathbf{k} \in K$$

The structure of the multibeam light wave $t(\mathbf{x})$ that is to be formed by the modan $M$ is depicted as schematic in Fig. 2.7.

Orthonormality property (1.3) can be rewritten as follows

$$(\psi_{\mathbf{p}}, \psi_{\mathbf{p}'}) = \delta_{\mathbf{pp}'} \tag{2.5}$$

Thus each partial beam contains selective by a single mode or a modal group $\{\psi_{\mathbf{p}}, \mathbf{p} \in L^{(\mathbf{k})}\}$ described by the equation

$$\xi^{(\mathbf{k})}(\mathbf{x}) = \sum_{\mathbf{p} \in L^{(\mathbf{k})}} \xi_{\mathbf{p}}^{(\mathbf{k})} \psi_{\mathbf{p}}(\mathbf{x}), \quad \mathbf{x} \in G, \quad \mathbf{k} \in K \tag{2.6}$$

The orthonormality property allows us to write down:

$$\left\|\xi^{(\mathbf{k})}\right\| = \left|\Xi^{(\mathbf{k})}\right|$$

The problem of selective excitation of the single isolated mode $\psi_{\mathbf{p}}$ in each branch is the special case of general consideration given above. The proper singlemodal multibeam modan forms the light wave

$$t(\mathbf{x}) = \sum_{\mathbf{p} \in K} \mu^{(\mathbf{p})} \psi_{\mathbf{p}}(\mathbf{x}) \exp[i 2\pi \mathbf{v}^{(\mathbf{p})} \mathbf{x}] \tag{2.7}$$

In this case, the number **k** of partial beam is same as the number **p** of the $p$-th mode $\psi_{\mathbf{p}}$ included in the singlemodal partial beam.

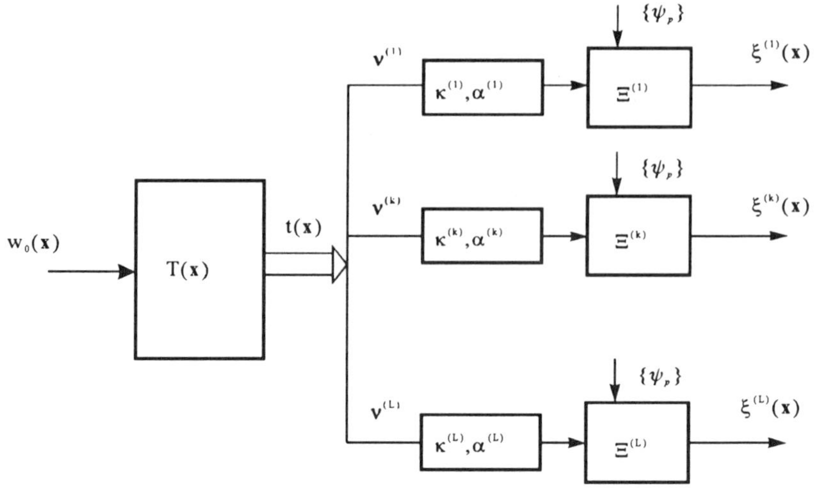


**Figure 2.7.** Schematic representation for the principle of multibeam modan operation.

$w_0(\mathbf{x})$-illuminating beam, $T(\mathbf{x})$-modan's complex transfer function, $t(\mathbf{x})$-beam with branches, reconstructed from modan, $\kappa^{(\mathbf{k})}$-relative power in one branch, $\alpha^{(\mathbf{k})}$-interbranch phase shifts, $\Xi^{(\mathbf{k})}$-vector of modal coefficients for one branch, $\xi^{(\mathbf{k})}(\mathbf{x})$-multimodal beam reconstructed in one branch of modan.

The set $L^{(\mathbf{p})}$ consists here of one index **p**. The index **p** runs the set $K$ comprising $L$ mode numbers.

Note in conclusion that multibeam modans are effective in following problems of fiber and laser optics: analysis of modal composition being performed simultaneously even for several groups of modes; selective excitation of various modal groups in fiber with the possibility of switching the group; simultaneous launching of radiation into a set of fibers; splitting of the fiber radiation associated with special mode selection; the mixing of modes accompanied by transform of modal composition.

## 2.4. GENERATING FUNCTIONS METHOD FOR THE SYNTHESIS OF MODANS

Development of both multibeam and multimodal modans is essentially hampered by the procedures of 2-D beam summation, which are described by Eqs.(2.6) or (2.2); this is especially the case for a large number of branches or a large number of modes in each branch. Computer generation procedure will involve summation of large files representing mode pixels. In addition, summation of the waves with near phases may result in a spasmodic increase in the magnitude of the modan's amplitude transmittance in separate points that, in turn, causes broadening the range of the amplitude transmittance variations and a decrease of the energy efficiency.

It is proposed [Golub, 1990c] to use the generating functions, which offer convenient methods for performing summation and provide smoothly varying complex transfer functions of modans for a large number of partial modal beams.

We define a function $\Pi$ as a generating function for a set of modes $\psi_{\mathbf{p}}(\mathbf{x})$ if it is an analytical function of the complex variables $\zeta$ and $\eta$ and have the following expansion into a $\zeta,\eta$ power series

$$\Pi(\mathbf{x},z,\zeta,\eta)=\sum_{\mathbf{p}}\pi_{\mathbf{p}}\gamma_{\mathbf{p}}\psi_{\mathbf{p}}(\mathbf{x})\,\zeta^{p}\eta^{l}, \tag{2.8}$$

Here, the indices $(p,l)$=**p** are non-negative and the eigenvalues $\gamma_{\mathbf{p}}$ are $z$-dependent. In view of the orthonormality of the modes $\psi_{\mathbf{p}}$, from Eq.(2.8) we can obtain the relationship

$$\int_G \Pi(\mathbf{x},z,\zeta,\eta)\psi_{\mathbf{p}}^{*}(\mathbf{x})\rho_0(\mathbf{x})d^2\mathbf{x}=\pi_{\mathbf{p}}\gamma_{\mathbf{p}}\zeta^{p}\eta^{l} \tag{2.9}$$

and a Parseval equality

$$E_\Pi=\int_G\left|\Pi(\mathbf{x},z;\zeta,\eta)\right|^2\rho_0(\mathbf{x})d^2\mathbf{x}=\sum_{\mathbf{p}}\left|\pi_{\mathbf{p}}\gamma_{\mathbf{p}}\right|^2\left|\zeta\right|^{2p}\left|\eta\right|^{2l} \tag{2.10}$$

We choose the complex variables $\zeta$ and $\eta$ in the form of the plane waves with the spatial frequencies $\mathbf{v}^{(10)}$ and $\mathbf{v}^{(01)}$ and initial phase $\alpha^{(10)}$ and $\alpha^{(01)}$

$$\zeta=s(\mathbf{x})=\left|\zeta\right|\exp(i2\pi\mathbf{v}^{(10)}\mathbf{x}+i\alpha^{(10)}),\quad\left|\zeta\right|<1 \tag{2.11}$$

Thus powers of $\zeta$ and $\eta$ will describe "multiple-spatial-frequency" plane waves having multiple tilts.

$$\eta=\eta(\mathbf{x})=\left|\eta\right|\exp(i2\pi\mathbf{v}^{(01)}\mathbf{x}+i\alpha^{(01)}),\quad\left|\eta\right|<1 \tag{2.12}$$

where $\left|\zeta\right|$ and $\left|\eta\right|$ are **x**-independent. The generating function, Eq.(2.8), yields then the singlemodal but multibeam expansion in the form of Eq.(2.7):

$$t(\mathbf{x})=\frac{1}{E_\Pi}\Pi(\mathbf{x},z,\zeta(\mathbf{x}),\eta(\mathbf{x}))=\sum_{\mathbf{p}}\sqrt{\kappa^{(\mathbf{p})}}\exp(i\alpha^{(\mathbf{p})})\psi_{\mathbf{p}}(\mathbf{x})\exp(i2\pi\mathbf{v}^{(\mathbf{p})}\mathbf{x}) \tag{2.13}$$

where

$$\kappa^{(\mathbf{p})}=(\left|\gamma_{\mathbf{p}}\pi_{\mathbf{p}}\right|\left|\zeta\right|^{p}\left|\eta\right|^{l})^2/E_\Pi,\quad\mathbf{p}=(p,l) \tag{2.14}$$

$$\alpha^{(\mathbf{p})} = \arg\gamma_{\mathbf{p}} + p\alpha^{(10)} + l\alpha^{(01)}, \quad \mathbf{v}^{(\mathbf{p})} = p\mathbf{v}^{(10)} + l\mathbf{v}^{(01)}, \qquad (2.15)$$

that describes a non-linear spatial modulation of the beam.

Note that the condition $\|t\| = 1$ is satisfied up on separating the branches and the factor $E_\Pi$, Eq.(2.10), introduced above can be represented as an integral of the fast-oscillating function

$$E_\Pi = \int_G |\Pi(\mathbf{x}, z, \zeta(\mathbf{x}), \eta(\mathbf{x}))|^2 \rho_0(\mathbf{x}) d^2\mathbf{x}. \qquad (2.16)$$

The relationship (2.1) is satisfied for the numbers $\kappa^{(\mathbf{p})}$. Subject to Eq.(2.10). As shown in section 2.1, function (2.13) describes a normalized field immediately behind the plane of the multibeam singlemodal modan with the interbranch power distribution in the form $\kappa^{(\mathbf{p})}$ and the phase distribution among the beams as $\alpha^{(\mathbf{p})}$.

A countable sequence, Eq.(2.15), of the spatial carrier frequencies $\mathbf{v}^{(\mathbf{p})}$ in the branches forms an equidistant discrete grid. When one needs to introduce symmetrical branches with the negative frequencies, Eq.(2.15), $p,l$=-1,-2,..., one can use a modified generating function

$$\Pi_*(\mathbf{x}, z; \zeta, \eta) = \Pi(\mathbf{x}, z; \zeta, \eta) + \Pi(\mathbf{x}, z; \zeta^*, \eta^*).$$

The modan with $t(\mathbf{x})$ in the form of Eq.(2.13) and with the multiple carrier frequencies, Eq.(2.15), proves to be especially convenient to perform modal filtration (see section 3.0). It is noteworthy that for $\mathbf{v}^{(10)} = \mathbf{v}^{(01)} = \mathbf{0}$ we obtain a singlebeam modan for a modal group representing the multimodal beam with the coefficients $\xi_{\mathbf{p}}^{(\mathbf{p})} = \mu^{(\mathbf{p})}$.

If a generative function is written in an analytical form, one can consider Eqs.(2.8) or (2.13) as convenient way for taking summation, provided that there are specific coefficients Eq.(2.9) of series. This is what determines a major role of the generating function.

It should be noted that a variety of generating functions is rather wide. So, beginning with the fixed generating function, Eq.(2.8), and multiplying it by the powers of $\zeta$ and $\eta$, and differentiating or integrating with respect to $\zeta$ and $\eta$, one can obtain an infinite set of generating functions that are different in the variations of the coefficients $\kappa^{(\mathbf{p})}$. Differentiation of Eq.(2.8), with respect to $\zeta$ and $\eta$, is seen to disperse the power in modes, whereas the integration of Eq.(2.8), with respect to $\zeta$ and $\eta$, concentrates the power in the finite number of modes and ensures rapid convergence of series (2.8) with respect to $\zeta$ and $\eta$. For example, if the procedure involves the multiplication of both sides of series (2.8) by $\zeta^{m_1}$ and $\eta^{j_1}$ followed by differentiation, with respect to $\zeta$ and $\eta$, being performed $p_1$ and $l_1$ times, respectively, with subsequent multiplication of the result by $\zeta^{m_2}$ and $\eta^{j_2}$

followed by the integration with respect to ζ and η, which is performed $p_2$ and $l_2$ times, respectively, the coefficients $\pi_{pl}$ are transformed according to the formula

$$\tilde{\pi}_{pl} = \pi_{pl} \frac{A^{l_1}_{l+j_1} \; A^{p_1}_{p+m_1}}{A^{l_2}_{l+l_2+j_2+j_1-l_1} \; A^{p_2}_{p+p_2+m_2+m_1-p_1}}, \tag{2.17}$$

where

$$A_n^m = n(n-1)\ldots(n-m+1)$$

If we interchange the operations of differentiation and integration, this yields us a different transformation

$$\tilde{\tilde{\pi}}_{pl} = \pi_{pl} \frac{A^{l_1}_{l+j_1+j_2+l_2} \; A^{p_1}_{p+m_1+m_2+p_2}}{A^{l_2}_{l+j_1+j_2} \; A^{p_2}_{p+m_2+p_2}} \tag{2.18}$$

By means of combination of the transformations (2.17) and (2.18) with various parameters, we can obtain a very wide variety of the power distributions between modes in different branches of modan. What is more, on summing up the various generating functions, one is able to obtain a group of more than one mode in each branch.

The inverse problem of choosing the generating function Π from the preset power distribution $\kappa^{(\mathbf{p})}$ between modes can be solved with the help of some purpose-oriented exhaustive search.

It seems expedient to make use of the table generating functions for the classical orthogonal polynomials and of the special functions to build up the assumed generating function of modes. So for the Gauss-Laguerre modes, the generating function can easily be produced from the well-known generating function for the Laguerre polynomials (see [Janke, 1960])

$$\frac{\exp\left(-\dfrac{\kappa\zeta}{1-\zeta}\right)}{(1-\zeta)^{l+1}} = \sum_{p=0}^{\infty} L_p^l(\kappa)\zeta^p \tag{2.19}$$

On multiplying both sides of Eq.(2.19) by the function

$$\kappa^{\frac{l}{2}} \exp\left(-\frac{\kappa}{2}\right) \exp(il\theta) \exp[i(l+1)\eta_0]$$

and substituting instead of $\alpha^{(10)}$ the value $2\eta_0 + \alpha^{(10)}$ with

$$\kappa = \frac{2r^2}{\sigma^2} = \frac{2\mathbf{x}^2}{\sigma^2}; \quad \eta_0 = \operatorname{arctg}\frac{z}{z_0}, \quad z_0 = \frac{\pi\sigma_0^2}{\lambda} \equiv b,$$

we really obtain the generating function for the series number $l$ of the Gauss-Laguerre modes represented by Eqs.(1.96) and (1.88). The final formula for Gauss-Laguerre generating function is

$$\Pi_l(\mathbf{x},z;\zeta)=\exp\left(-\frac{r^2}{\sigma^2}\cdot\frac{1+\zeta}{1-\zeta}\right)\left(\frac{\sqrt{2}r}{\sigma}\right)^l\frac{1}{(1-\zeta)^{l+1}}\exp[il\theta+i(l+1)\eta_0], \quad (2.20)$$

where σ is the radius of a principal mode, $r$ and θ are the polar radius and the angle of the point **x** in the modan plane, respectively, and the variable η is chosen equal to one, subject to a fixed value $l$. In this case

$$\pi_{pl}=\frac{1}{E_{pl}}=\sigma\sqrt{\frac{\pi(p+l)!}{2p!}},$$

and $\kappa^{(pl)}$ is specified by Eqs.(2.14) and (2.10). If we perform an additional multiplication of both sides of Eq.(2.20) by $\eta^l$, which is followed by summation within $l=\overline{l_1,l_2}$, this gives us the generating function for the series of the $\Delta l=l_2+l_1+1$ Gauss-Laguerre modes equal to

$$\Pi_{l_1l_2}(\mathbf{x},z;\zeta,\eta)=\Pi_{l_1}(\mathbf{x},z;\zeta)\frac{1-\tilde{\Pi}^{\Delta l}}{1-\tilde{\Pi}}, \quad (2.21)$$

where

$$\tilde{\Pi}=\frac{\sqrt{2}r}{\sigma}\cdot\frac{\eta}{1-\zeta}\exp(i\theta+i\eta_0) \quad (2.22)$$

In particular, for $l_1=0$ and $l_2\to\infty$ and subject to the condition

$$\frac{\sqrt{2}\,r|\eta|}{\sigma|1-\zeta|}<1,$$

we obtain the generating function for the entire set of the Gauss-Laguerre modes

$$\Pi(\mathbf{x},z;\zeta,\eta)=\frac{\Pi_0(\mathbf{x},z;\zeta)}{1-\tilde{\Pi}}, \quad (2.23)$$

where from Eq.(2.20) we find that

$$\Pi_0(\mathbf{x},z;\zeta)=\frac{\exp(i\eta_0)}{1-\zeta}\exp\left(-\frac{r^2}{\sigma^2}\cdot\frac{1+\zeta}{1-\zeta}\right) \quad (2.24)$$

The use of Eqs.(2.22) and (2.23) makes it possible to write Eqs.(2.20) and (2.21) in the form

$$\Pi_l(\mathbf{x},z;\zeta)=\left(\frac{\sqrt{2}r}{\sigma}\right)^l(1-\zeta)^{-l}\exp[il(\theta+\eta_0)]\Pi_0(\mathbf{x},z;\zeta) \quad (2.25)$$

$$\Pi_{l_1 l_2}(\mathbf{x},z;\zeta,\eta)=\frac{\Pi_{l_1}(\mathbf{x},z;\zeta)\Pi(\mathbf{x},z;\zeta,\eta)}{\Pi_0(\mathbf{x},z;\zeta)}\cdot[1-\tilde{\Pi}^{\Delta l}] \tag{2.26}$$

The function $T(\mathbf{x})$ of the complex transfer of the Gauss-Laguerre multibeam modan can be obtained from Eqs.(2.20), (2.23), and (2.24) using the substitution into Eqs.(2.23) through (2.26) of the functions (2.11) and (2.12) instead of the parameters $\zeta$ and $\eta$, and with $|\eta|$ and $|\zeta|$ taken as parameters, which are less than one.

***Example.***

Let us consider in more details the generating function for the radial symmetrical ($l=0$) Gauss-Laguerre modes at $z=0$ and $\mathbf{v}^{(10)}=(\nu^{(10)},0)$. Table 2.1 and Fig. 2.8 summarize the $\kappa^{(p,0)}$ power distribution in the modan's branches for the Gauss-Laguerre modes with $l=0$ and $z=0$ and for different $p$ when using generating function (2.24). The factor $|\zeta|$ is chosen as a parameter.

One can see from Fig. 2.8 that with the increasing parameter $|\zeta|$, the power disperses in a larger number of modes and, vice versa, the decreasing $|\zeta|$ results in the concentration of power in the principal mode. In the derivatives of the generating function (Figs. 2.9 through 2.11 and Table 2.2) the power maximum is displaced to the first or to the higher order modes. Multiplication on the power performed prior to differentiation yields a width broadening for a curve of the power distribution (Fig. 2.11).

The module and phase of the generating function $\Pi_0$ are specified in accordance with Eqs.(2.24) and (2.11) as follows

$$|\Pi_0(\mathbf{x},0,\zeta)|=\frac{1}{|1-\zeta|}\exp\left\{-\frac{r^2}{\sigma^2}\cdot\frac{1-|\zeta|^2}{|1-\zeta|^2}\right\},\quad \mathbf{r}=|\mathbf{x}|$$

$$\arg\Pi_0(\mathbf{x},0,\zeta)=\mathrm{arctg}\left(\frac{|\zeta|\sin\varphi}{1-|\zeta|\cos\varphi}\right)-\frac{r^2}{\sigma^2}\cdot\frac{2|\zeta|\sin\varphi}{|1-\zeta|^2}$$

where

$$|1-\zeta|^2=1-2|\zeta|\cos\varphi+|\zeta|^2$$

$$\varphi=2\pi\mathbf{v}^{(10)}\mathbf{x}+\alpha^{(10)}$$

The $\varphi$ phase variation in the range from 0 to $2\pi$ corresponds to one period of the oscillations of $\Pi_0$ function. Two envelope lines of $\Pi_0$ function correspond to the values:

$$\varphi=0,\quad |1-\zeta|=1-|\zeta|;\qquad \varphi=\pi;\quad |1-\zeta|=1+|\zeta|$$

**Table 2.1**
**The values of interbranch modal power distribution for the generating function**

| $\lvert\zeta\rvert$ | $\kappa^{(0,0)}$ | $\kappa^{(1,0)}$ | $\kappa^{(2,0)}$ | $\kappa^{(3,0)}$ | $\kappa^{(4,0)}$ | $\kappa^{(5,0)}$ | $\kappa^{(6,0)}$ |
|---|---|---|---|---|---|---|---|
| 0.25 | 0.937 | 0.058 | 0.004 | <0.001 | <0.001 | <0.001 | <0.001 |
| 0.33 | 0.888 | 0.098 | 0.011 | 0.001 | <0.001 | <0.001 | <0.001 |
| 0.50 | 0.750 | 0.188 | 0.047 | 0.012 | 0.003 | 0.001 | <0.001 |
| 0.75 | 0.438 | 0.246 | 0.138 | 0.078 | 0.044 | 0.025 | 0.014 |

Modified from Golub M.A., Sisakyan I.N., Soifer V.A., Computer generated optical elements for fiber's mode selection and launching, Proceedings SPIE, OE/Fibers'90, 1365, 156, 1990

**Table 2.2**
**The values of interbranch modal power distribution for the derivative of generating function**

| $\lvert\zeta\rvert$ | $\kappa^{(1,0)}$ | $\kappa^{(2,0)}$ | $\kappa^{(3,0)}$ | $\kappa^{(4,0)}$ | $\kappa^{(5,0)}$ | $\kappa^{(6,0)}$ | $\kappa^{(7,0)}$ | $\kappa^{(8,0)}$ |
|---|---|---|---|---|---|---|---|---|
| 0.25 | 0.776 | 0.194 | 0.027 | <0.001 | <0.001 | <0.001 | <0.001 | <0.001 |
| 0.33 | 0.632 | 0.281 | 0.070 | 0.014 | <0.001 | <0.001 | <0.001 | <0.001 |
| 0.5 | 0.338 | 0.388 | 0.190 | 0.084 | 0.033 | 0.019 | <0.001 | <0.001 |
| 0.75 | 0.054 | 0.121 | 0.153 | 0.153 | 0.134 | 0.109 | 0.061 | 0.061 |

Modified from Golub M.A., Sisakyan I.N., Soifer V.A., Computer generated optical elements for fiber's mode selection and launching, Proceedings SPIE, OE/Fibers'90, 1365, 156, 1990 .

Figures 2.12 through 2.15 present the modulus and phase of the function $\Pi_0$ versus the coordinate x of the vector **x**=$(x,y)$, for $y=0$ and $z=0$ and with the different values of the parameter $\lvert\zeta\rvert$ (solid line for $\lvert\zeta\rvert=0.25$, dashed line for $\lvert\zeta\rvert=0.5$). The normalization of $\Pi_0$ was performed by the constant

$$\|\Pi_0\| = \sigma\sqrt{\frac{\pi}{1-\lvert\zeta\rvert^2}}$$

If $\lvert\zeta\rvert=0.25$, the major portion of power is concentrated in the principal mode (0,0), and the curves shown in Figs. 2.12a and 2.13a represent the modulation of $\Pi_0$ with a varying spatial frequency. When $\lvert\zeta\rvert=0.5$, a substantial portion of power is redistributed into the mode (1,0), and the in-phase superposition of the modes with $\alpha^{(10)}=0$ leads to a rather pronounced central peak (see dashed lines in Figs. 2.12a and 2.13a).

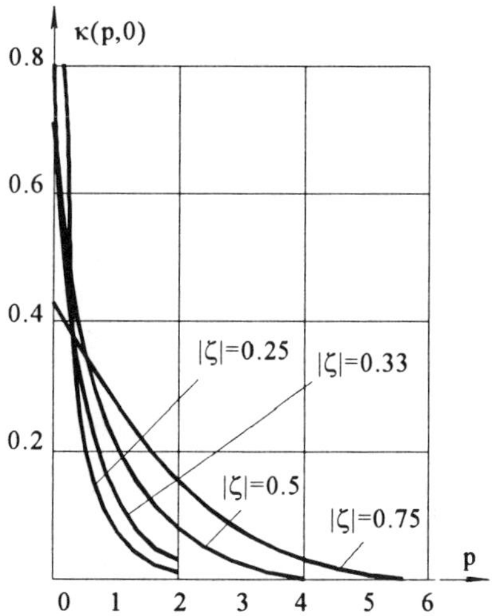


**Figure 2.8.** Interbranch modal power $\kappa^{(p,0)}$distribution for the Gauss-Laguerre generating function $\Pi_0$. p - radial mode number. (Modified from Golub M.A., Sisakyan I.N., Soifer V.A., Computer generated optical elements for fiber's mode selection and launching, Proceedings SPIE, OE/Fibers'90, 1365, 156, 1990).

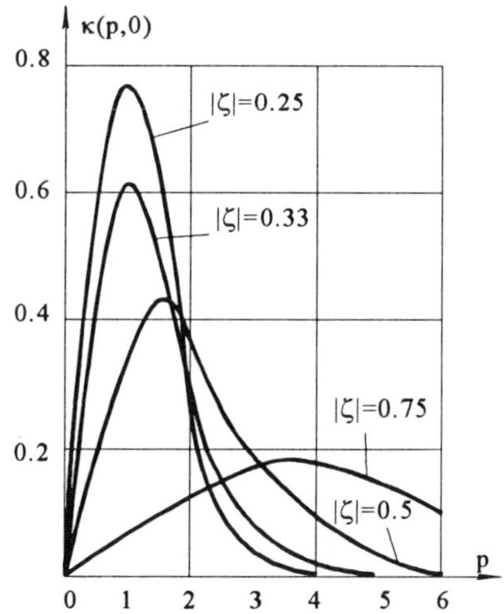


**Figure 2.9.** Interbranch modal power $\kappa^{(p,0)}$ distribution after the first-order differentiation of the Gauss-Laguerre generating function $\Pi_0$. $p$ - radial mode number. (Modified from Golub M.A., Sisakyan I.N., Soifer V.A., Computer generated optical elements for fiber's mode selection and launching, *Proceedings SPIE, OE/Fibers'90*, 1365, 156, 1990).

When choosing a superposition, which is performed in anti-phase (see Fig. 2.14), with $\alpha^{(10)}=\pi$, the central peak disappears and the range of variations of the generating function modulus essentially decreases. For $\alpha^{(10)}=\pi/2$, (see Fig. 2.15) there arises some asymmetry of the generating function.

With reference to the Bessel modes we can use the expansion described by [Janke, 1960].

$$\exp\left[\frac{\kappa}{2}\left(\zeta-\frac{1}{\zeta}\right)\right]=J_0(\kappa)+\sum_{l=1}\left[\zeta^l+(-\zeta)^{-l}\right]J_l(\kappa)$$

Using the series described by [Janke, 1960] we can in a similar fashion build up the generating function for the Gauss-Hermite modes

$$\exp(2\kappa\zeta-\zeta^2)=\sum_{p=0}^{\infty}H_p(\kappa)\frac{\zeta^p}{p!},$$

where $Hp$ is the Hermite polynomial. Multiplying both sides of the equality by $\exp\left(-\chi^2/2\right)$ gives us

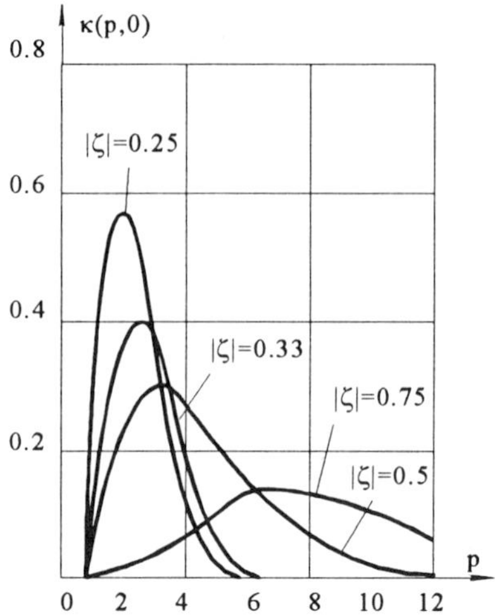


**Figure 2.10.** Interbranch modal power $\kappa^{(p,0)}$ distribution after the second-order differentiation of Gauss-Laguerre generating function $\Pi_0$. $p$ - radial mode number.

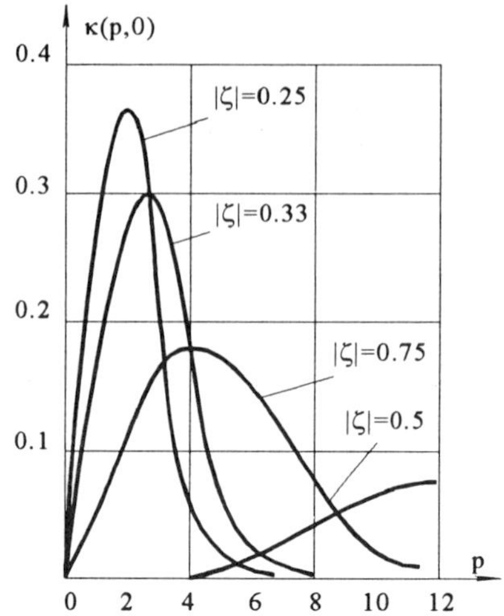


**Figure 2.11.** Interbranch modal power $\kappa^{(p,0)}$ distribution for generating function $\frac{\partial^5}{\partial\zeta^5}\left(\zeta^5\Pi_0\right)$, $\Pi_0$ - generating function of Gauss-Laguerre modes, $p$ - radial mode number.

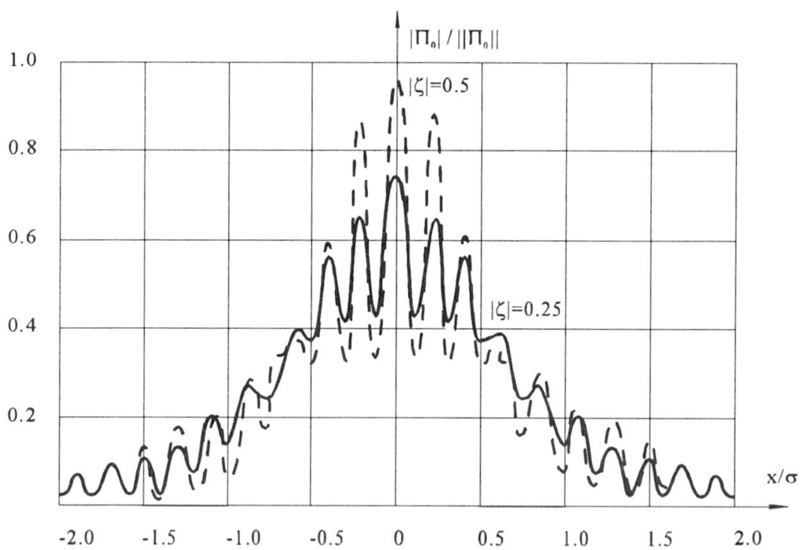


**Figure 2.12.** Normalized modulus of the generating function $\Pi_0$ for Gauss-Laguerre modes. $|\zeta|$-parameter of generating function; σ-radius of the principal Gaussian mode at the level of exp(-2) intensity fall; initial phase $\alpha^{(10)} = 0$, carrier frequency $\nu^{(10)} = 5$ lines/mm.

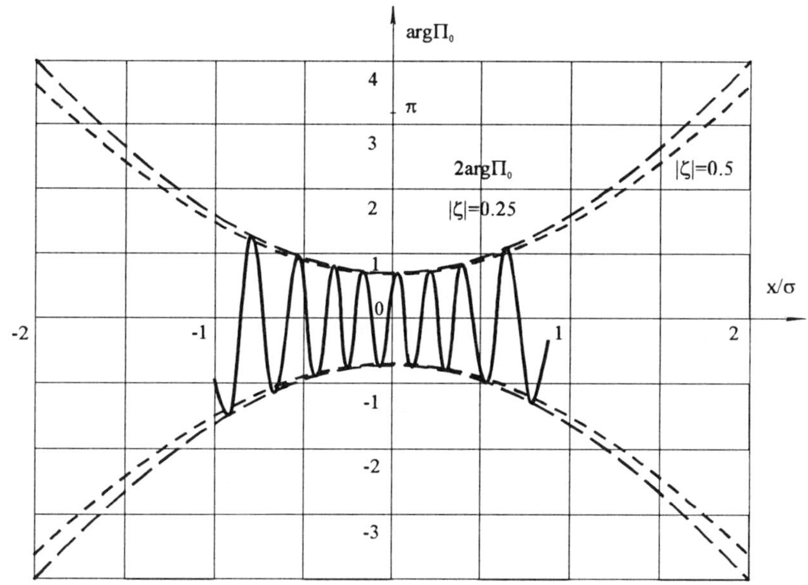


**Figure 2.13.** Phase of the generating function $\Pi_0$ for Gauss-Laguerre modes. Dashed curve shows the envelope line of phase. $|\zeta|$-parameter of generating function; initial phase $\alpha^{(10)} = 0$, carrier frequency $\nu^{(10)} = 5$ lines/mm; σ-radius of the principal Gaussian mode at the level of exp(-2) intensity fall.

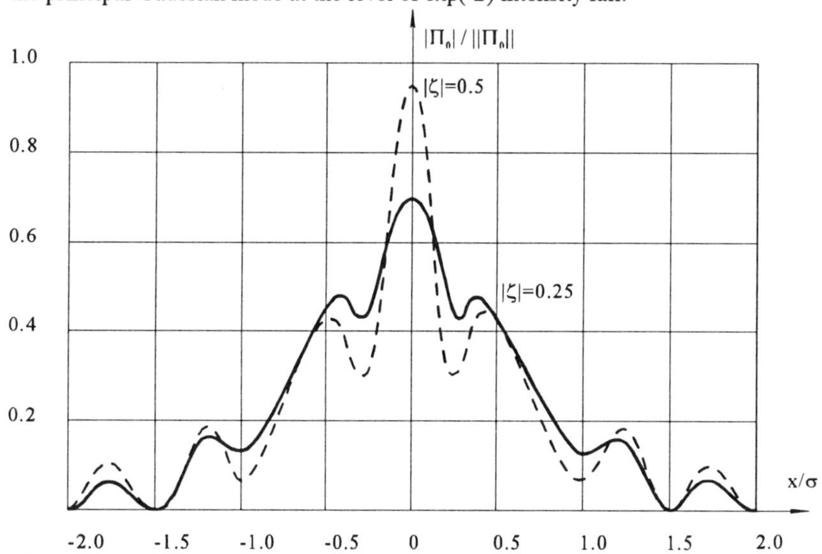


**Figure 2.14.** Normalized modulus of the generating function $\Pi_0$ for Gauss-Laguerre modes. $|\zeta|$-parameter of generating function; σ-radius of principal Gaussian mode at the level of exp(-2) intensity fall; initial phase $\alpha^{(10)} = 0$, carrier frequency $\nu^{(10)} = 2$ lines/mm.

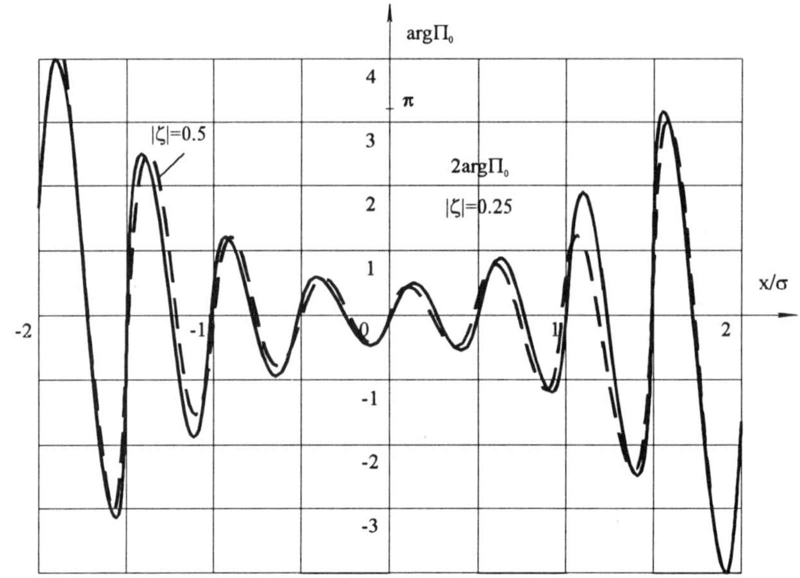


**Figure 2.15.** Phase of the generating function $\Pi_0$ for Gauss-Laguerre modes. Dashed curve shows the envelope line of phase. $|\zeta|$-parameter of generating function; Initial phase $\alpha^{(10)} = 0$, carrier frequency $\nu^{(10)} = 2$ lines/mm; σ-radius of principal Gaussian mode at the level of exp(-2) intensity fall.

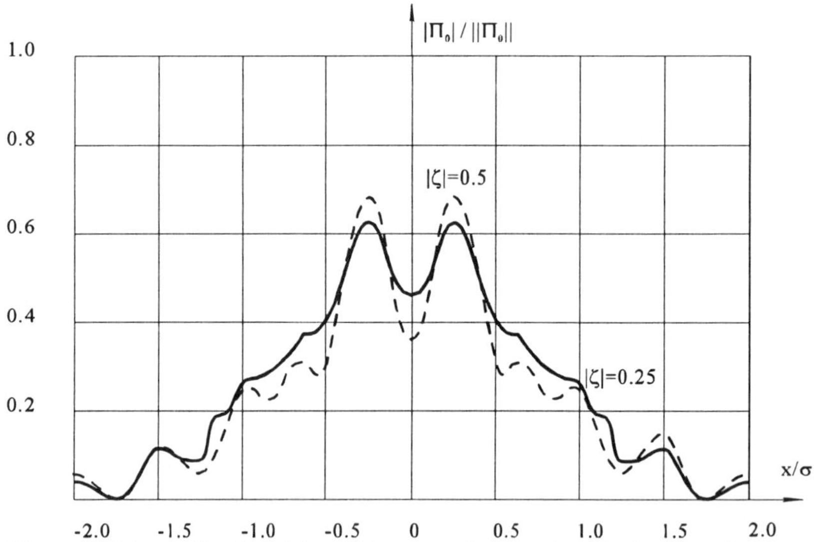


**Figure 2.16.** Normalized modulus of the generating function $\Pi_0$ for Gauss-Laguerre modes. $|\zeta|$-parameter of generating function; δ-radius of principal Gaussian mode at the level of exp(-2) intensity fall; initial phase $\alpha^{(10)} = \pi$, carrier frequency $\nu^{(10)} = 2$ lines/mm.

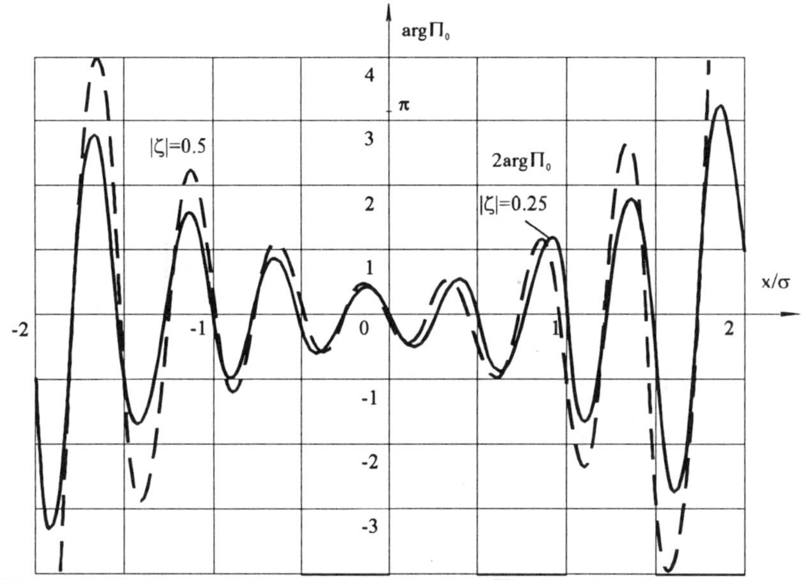


**Figure 2.17.** Phase of the generating function $\Pi_0$ for Gauss-Laguerre modes. Dashed curves show the envelope line of phase. $|\zeta|$-parameter of generating function; initial phase $\alpha^{(10)} = \pi$, carrier frequency $\nu^{(10)} = 2$ lines/mm; σ-radius of principal Gaussian mode at the level of exp(-2) intensity fall.

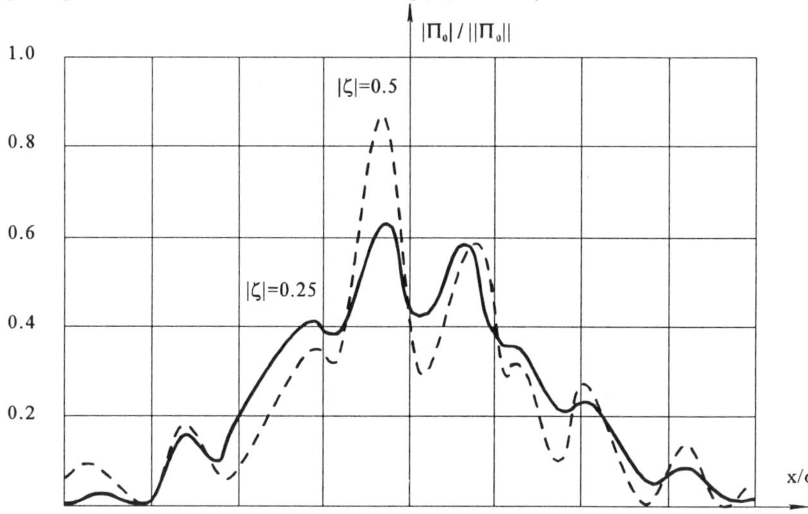


**Figure. 2.18.** Normalized modulus of the generating function $\Pi_0$ for Gauss-Laguerre. $|\zeta|$-parameter of generating function; σ-radius of principal Gaussian mode at the level of exp(-2) intensity fall; initial phase $\alpha^{(10)} = \pi/2$, carrier frequency $\nu^{(10)} = 2$ lines/mm.

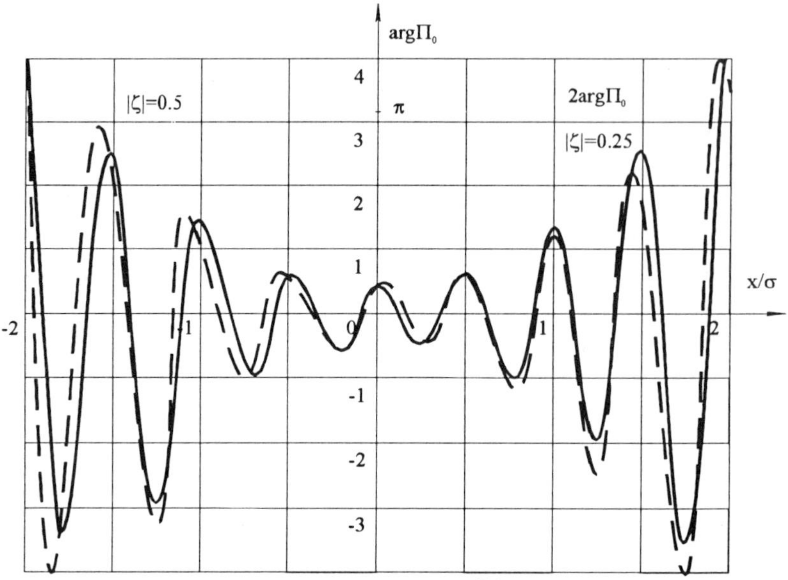


**Figure 2.19.** Phase of the generating function $\Pi_0$ for Gauss-Laguerre modes. Dashed curves show the envelope line of phase. $|\xi|$-parameter of generating function; initial phase $\alpha^{(10)} = \pi/2$, carrier frequency $\nu^{(10)} = 2$ lines/mm; σ-radius of principal Gaussian mode at the level of exp(-2) intensity fall.

$$\exp(2\kappa\zeta - \zeta^2 - \frac{\kappa^2}{2}) = \sum_{p=0}^{\infty} H_p(\kappa)\exp\left(-\frac{\kappa^2}{2}\right)\frac{1}{p!}\zeta^p \tag{2.27}$$

and similar to this

$$\exp(2\tilde{\kappa}\eta - \eta^2 - \frac{\tilde{\kappa}^2}{2}) = \sum_{l=0}^{\infty} H_l(\tilde{\kappa})\exp\left(-\frac{\tilde{\kappa}^2}{2}\right)\frac{1}{l!}\eta^l \tag{2.28}$$

On multiplying Eq.(2.27) by Eq.(2.28), assuming

$$\alpha^{(10)} = \eta_0, \quad \alpha^{(01)} = \eta_0, \quad \kappa = \frac{\sqrt{2}x}{\sigma}, \quad \tilde{\kappa} = \frac{\sqrt{2}y}{\sigma}$$

and making use of definition (1.108) for the Gauss-Hermite modes, we obtain the generating function for the complete set of the Gauss-Hermite modes in the form

$$\Pi(\mathbf{x}, z; \zeta, \eta) = \exp\left(\frac{2\sqrt{2}(x\xi + y\eta)}{\sigma} - \zeta^2 - \eta^2 - \frac{x^2 + y^2}{\sigma^2} + i\eta_0\right) \tag{2.29}$$

In this case

$$\pi_{pl} = \frac{1}{E_{pl} p!l!} = \sigma \sqrt{\frac{\pi}{2} \cdot \frac{2^{p+l}}{p!l!}} \tag{2.30}$$

and the value $\kappa^{(p,l)}$ can be calculated from Eqs.(2.14) and (2.10). Note that $\pi_{pl}$ is separable with respect to indices. On substituting the ζ and η values from Eqs.(2.11) and (2.12) into Eq.(2.29) we can obtain the function of the complex transfer for the multibeam Gauss-Hermite modan.

Let us consider in conclusion another important technique that allows the generating function to be build up. This technique makes use of the fact (see [Perelomov, 1986], [Krivoshlykov, 1987]) that a generating function for the mode is each the ray function being the analogs of the coherent states that are considered in quantum mechanics for the set of the stationary states (modes). In this case the following expansion takes place (see [Perelomov, 1986]):

$$\Pi(\mathbf{x}, z; \zeta, \eta) = \exp\left( -\frac{|\alpha|^2 + |\beta|^2}{2} \right) \cdot \sum_{p,l=0}^{\infty} \frac{\alpha^p \beta^l}{\sqrt{p!l!}} \psi_{pl}(\mathbf{x}),$$

where α and β are the parameters of the ray functions, which are similar to $\zeta$ and $\eta$ in Eq.(2.8).

In particular, the well-known relationship [Landau, 1974] exists for the coherent states corresponding to the Gauss-Hermite modes (1.108) in free space

$$\Pi(\mathbf{x}, z; \alpha, \beta) = \Pi(x, z; \alpha) \tilde{\Pi}(y, z; \beta) \tag{2.31}$$

where

$$\Pi(x, z; \alpha) = \frac{1}{(2\pi)^{1/4}} \cdot \frac{1}{\sqrt{\sigma/2}} \exp\left( -i \frac{k}{2} \bar{x} \bar{p}_{x_0} - i \frac{\eta_0}{2} + i \frac{k}{2} z \right) \times$$
$$\times \exp\left[ -\frac{(x - \bar{x})^2}{\sigma^2} + ik \bar{p}_{x_0} x \right],$$

where the parameters σ, R, and $\eta_0$ are specified by Eqs.(1.88), (1.91), $\bar{x}_0$ is an average initial coordinate, and $\bar{p}_{\mathbf{x0}}$ is an average tilt, both being related to the complex parameter α by the equations

$$\mathrm{Im}\,\alpha = \frac{1}{\sigma} \left[ z_0 \bar{p}_{x_0} - (\bar{x}_0 - z \bar{p}_{x_0}) \frac{z}{2z_0} \right]$$

$$\mathrm{Re}\,\alpha = \frac{1}{\sigma} (\bar{x}_0 + z \bar{p}_{x_0}); \quad \bar{x} = \bar{x}_0 + z \bar{p}_{x_0}$$

The function $\tilde{\Pi}(y, z; \beta)$ is written in a similar fashion.

The function of the modan complex transfer is deduced from Eqs.(2.11) and (2.12) by means of varying the parameters α and β with respect to $x$ at $\alpha = \zeta$ and with $\beta = \eta$.

## 2.5. DIGITAL HOLOGRAPHY METHODS FOR GENERATION OF MODANS

### A. Coding-method problem in digital holography

The modans are to ensure a complex-valued transfer functions since they form complex (amplitude-phase) mode beams. At the same time, the energy efficiency considerations require that the modans should be realized as the phase-only optical elements. The problem of recording the complex light field on the phase-only medium is well known in digital holography and reviewed in [Huang, 1971; Yaroslavskii 1980; Dallas, 1986; Wyrowski, 1991]. The appropriately synthesized thin holograms have a phase microrelief on the plane substrate. Thus, in terms of digital holography, development of the modan is the development of the phase synthesized hologram. Realization of the phase diffractive elements by means of digital holography involves the following stages: the computation of the complex (amplitude-phase) transfer function, its transformation into the phase-only function, generation of masks, and creation of the phase microrelief (see Fig. 2.20).

The process of recording the amplitude-phase transfer function onto the synthesized hologram involves substantial difficulties. The multilayer elements of a "ROACH" type [Chu, 1973][Chu, 1974] are too cumbersome and absorb substantial portion of the illuminating beam energy. "Tandem" optical components [Bartelt, 1985] as well as "dual-phase" optical components [Mait, 1987] have higher energy efficiency, but they imply the use of the complicated two-stage optical systems.

Special coding methods are available in digital holography. The methods are aimed at creating the phase-only holograms providing the required amplitude-and-phase transfer function. A method of kinoform [Lesem, 1969] complemented by the special methods for making the amplitude more uniform [Chu, 1972; Akaev, 1977] are powerful tools in phase-only holography. The digital holography methods for introducing a random or deterministic additional phase into the reconstructed image [ Dallas, 1986] are interesting for intensity-type images but are of no use upon forming the complex(amplitude-phase) object. The most appropriate approach for coding the modans is the diffraction redistribution of a beam energy. Such an approach has found use in the method of detour phase with the binary-phase cells [Brown, 1966; Bergeron 1993; Campos, 1993], in the method of parity elements [Chu, 1972] and in the Kirk-Jones method [Kirk, 1971] as well as in the theory of the phased antenna arrays [Kashin, 1970].

In this section we introduce and also substantiate a general method to perform diffractive coding of the complex transfer function into the form of the phase-only function. The foregoing methods of digital holography are shown to be specific cases.

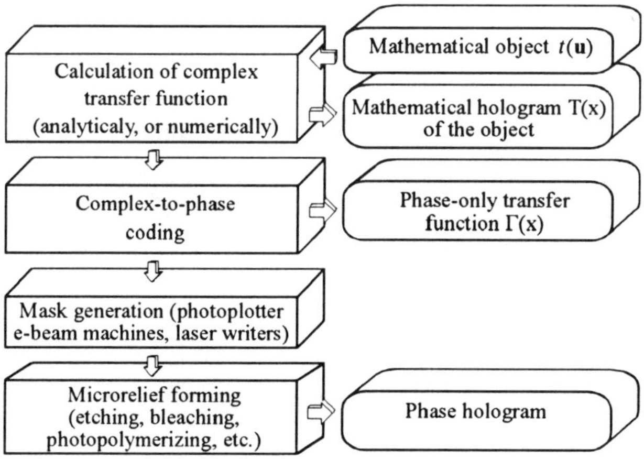


**Figure 2.20.** Stages of computer-aided generation of phase holograms.

Let us consider a modan as a thin digital hologram on the plane substrate. The orthogonal Cartesian coordinates $\mathbf{x}=(x,y)$, with $z$-axis perpendicular to $\mathbf{x}$ and with the light aperture defined in the domain $G$ of the $\mathbf{x}$-variables (see Figs. 2.21), are introduced.

Let us request the beam $t(\mathbf{x})$ to be reconstructed directly behind the holograms plane upon the illuminating beam $w_0(\mathbf{x})$. Thus, a "mathematical" hologram should perform point-by-point transform of light field with the (complex amplitude-phase) transfer function $T(\mathbf{x})$ proportional to $t(\mathbf{x})/w_0(\mathbf{x})$. Let us denote a coefficient of proportionality as $c\sqrt{E_0/E}$, where

$$E_0 = \int_G |w_0(\mathbf{x})|^2 d^2\mathbf{x}, \quad E = \int_G |t(\mathbf{x})|^2 d^2\mathbf{x}$$

are the total powers of the $w$-and $t$-fields, respectively. Thus, required complex transfer is given be the formula

$$T(\mathbf{x}) = c\sqrt{E_0/E}\cdot\frac{t(\mathbf{x})}{w_0(\mathbf{x})} \tag{2.32}$$

The constant $c$ is chosen from the condition of normalizing the values $T(\mathbf{x})$ up to the interval:

$$|T(\mathbf{x})| \leq 1$$

The equation for $c$ constant can be written as follows:

$$c^{-1} = \max_{\mathbf{x}\in G}\left|\frac{t(\mathbf{x})/\sqrt{E}}{w_0(\mathbf{x})/\sqrt{E_0}}\right| \tag{2.33}$$

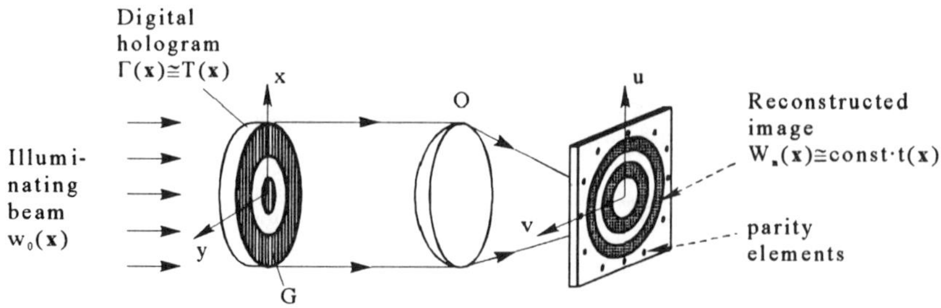


**Figure 2.21.** Optical image reconstruction from phase-only hologram. Fourier-lens $O$ converts plane-wave diffraction orders into the points of output plane. Parity elements occur due to the coding of the complex transfer function into the phase-only one.

In terms of mathematics, the coding represents the change over from the calculated function of the complex transfer

$$T(\mathbf{x}) = |T(\mathbf{x})| \exp(i\varphi_T(\mathbf{x})), \quad |T(\mathbf{x})| \le 1$$

to the "coded" phase-only transfer function

$$\Gamma = \exp(i\varphi) \tag{2.34}$$

were $\varphi = \varphi(\mathbf{x})$ is the coded phase.

**B. Encoding procedures based on modulation theory**

The idea of coding from complex to phase-only transfer is allowing some "parity" or auxiliary elements in the reconstructed image (see Fig. 2.21). These parity elements correspond to parasite diffraction orders and are indeed separated from the main operating order. They are only redistributing the beam power and reduce respectively the power of main image.

A formal description of "coded" phase holograms is based on the phase-modulation theory, well known in radio-engineering. Of course, an optical application has peculiarities related with 2-D spatial coordinates present instead of the temporal 1-D variable. For this purpose we choose a spatial carrier function that is some basic phase diffraction grating, or another periodic structure.

One of the cases for spatial carrier structures is a 1-D phase grating with a period of $d_0$ and a spatial frequency of $v_1 = 1/d_0$. Such a 1-D grating can be sinusoidal, binary-stepped, sawtooth-type (see Fig. 2.22) or the other type described by a dimensionless function $q_0(t)$ taking values on the [-0.5, 0.5] interval and periodic with the period 1 with respect to the dimensionless variable $t$. We can call $q_0(t)$ function as groove- shape function. One can simply superimpose two crossed 1-D gratings and obtain 2-D gratings as spatial carriers for coding.

Other special types of 2-D periodic structures are really wide spread in digital holography [Huang, 1971; Yaroslavskii, 1986]. They are called "coding cells" and include Lohmann's cells [Brown, 1966] and Ichioka's cells [Huang, 1971] displayed in Figs. 2.23, 2.24. The cell with the size $d_x, d_y$ will be characterized by the 2-D groove-shape function $q(\mathbf{t}) \equiv q(t_1, t_2)$, $\mathbf{t} = (t_1, t_2)$ with values on the [-0.5, 0.5] interval. The function $q$ is assumed to be periodic with the periods 1 with respect to each of $t_1, t_2$ variables. One period of the $q(t_1, t_2)$ function on the $t_1, t_2$ plane occupies the unit square $0 \le t_1 \le 1, 0 \le t_2 \le 1$ and describes single coding cell.

Maximum height of microrelief in the coding cell also can be modulated by multiplying $q$ by factor $a$ that is slowly varying from cell to cell. Thus $aq(t_1, t_2)$ term will be present in "coded" phase. Furthermore, the cell can be shifted from the regular position by using the function $q(t_1 - \alpha_1, t_2 - \alpha_2)$ where detour parameters $\boldsymbol{\alpha} = (\alpha_1, \alpha_2)$ are modulated and thus slowly varying from cell to cell.

The function $q(\mathbf{t})$ and the structure of the cell may depend on some additional parameters designated as $\boldsymbol{\tau}$ vector. Window size $\tau$ could be considered as an example of additional parameters. The parameters $\boldsymbol{\tau}$ can be slowly varying (modulated) within the plane of optical element. We are introducing $\boldsymbol{\tau}$ parameters into the function $q$ as $q(\mathbf{t}; \boldsymbol{\tau})$.

It is clear that the slow variations of any of the parameters listed above change the diffraction conditions and thus result in a beam power redistribution between diffraction orders, that is equivalent to directing light into parity elements (see Fig. 2.21).

Thus, the main operating diffraction order uses different parts of power from different coding cells. This effect is equivalent to introduction of implicit amplitude transfer by phase-only structure. Such an effect is the core of coding mechanism.

It must be emphasized once more that all modulated parameters must vary slowly so that their spatial featuresize $\Delta x_T, \Delta y_T$ is much more than the period $d_0$ (or $d_x$, $d_y$) of spatial carrier grating (see Fig. 2.25).

In this case diffraction orders are angular separated and modulation is simply described. In other words the spatial frequencies of basic diffraction grating must be high enough in comparison with the $\Delta x_T^{-1}, \Delta y_T^{-1}$ values.

Let the basic diffraction grating have the carrier spatial frequencies $\hat{S}$, and the maximum height of the groove equal to $\varphi_{\max}$ (in terms of the phase delay). Here $\hat{S}$ is the matrix

$$\hat{S} = \begin{pmatrix} S_{11} & S_{12} \\ S_{21} & S_{22} \end{pmatrix}$$

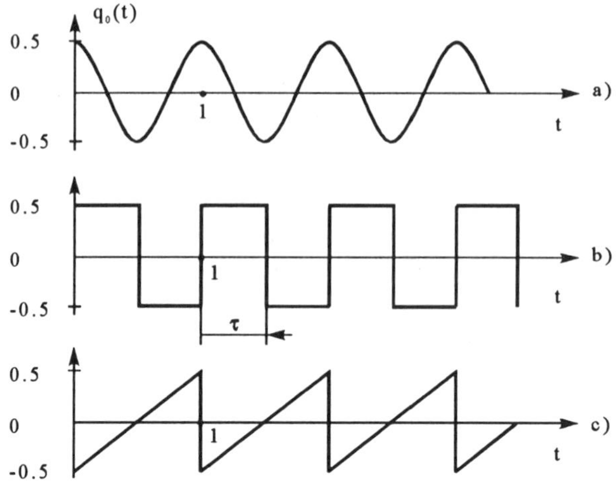


**Figure 2.22.** Typical shapes of 1-D basic diffraction gratings (carrier function): a) cosinusoidal; b) rectangular-pulsed; c) sawtooth.

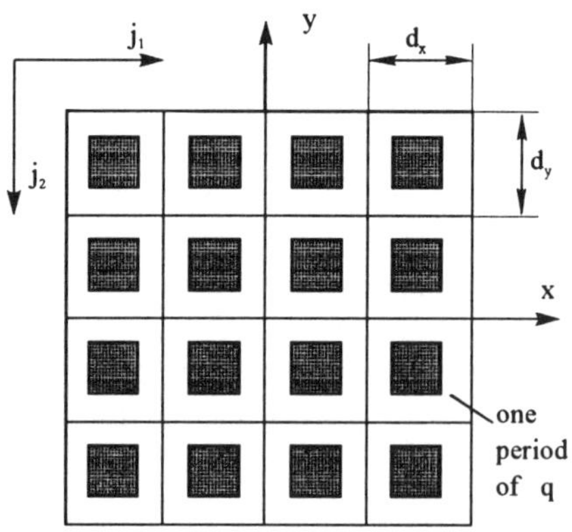


**Figure 2.23.** Basic diffraction grating for coding method by A.W. Lohmann, $d_x$, $d_y$-sizes of cell, $\nu_{1x}, \nu_{1y}$-carrier spatial frequencies.

The eigenvalues $\nu_{1x}, \nu_{1y}$ of the $\hat{S}$ matrix correspond to spatial frequencies that we have with respect to $x$ and $y$ axis. Thus sizes of single coding cell will be $d_x = \dfrac{1}{|\nu_{1x}|}$; $d_y = \dfrac{1}{|\nu_{1y}|}$

Now we are building some generalization of the Kirk-Jones method [Kirk, 1971]. Diffractive optical element has a spatially modulated phase function [Adzhalov, 1990; Golub, 1989] in the form

$$\varphi(\mathbf{u}) = b(\mathbf{x}) + \varphi_{\max} \beta a(\mathbf{x})\, q[\hat{S}\mathbf{x} + \alpha(\mathbf{x}); \tau] \tag{2.35}$$

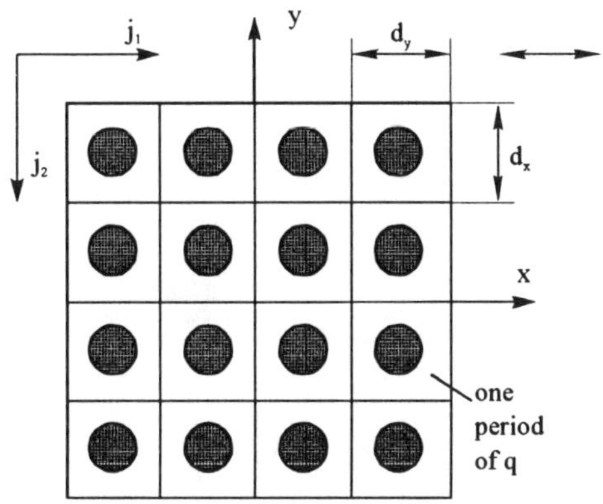


**Figure 2.24.** Basic diffraction grating for coding method by Y. Ichioka. $d_x$, $d_y$-sizes of cell, $\nu_{1x}, \nu_{1y}$-carrier spatial frequencies.

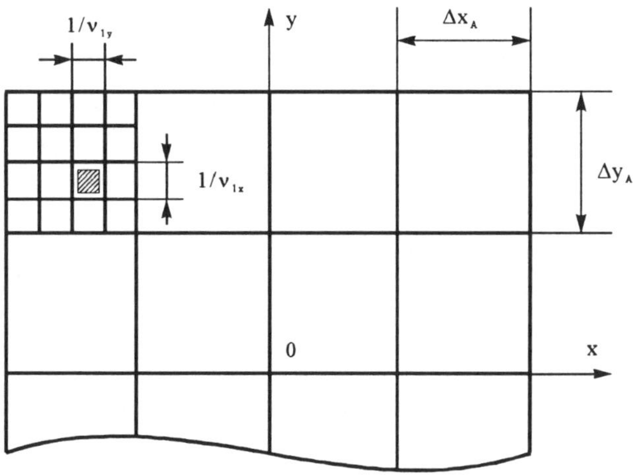


**Figure 2.25.** Coding cells with sizes $d_x = 1/|\nu_{1x}|$, $d_y = 1/|\nu_{1y}|$. Parameters $\Delta x_T, \Delta y_T$- characterize featuresize interval for variation of required complex transfer $T$ ; $\Delta x_T >> d_x, \Delta y_T >> d_y$ because the $T$ function varies slowly as compared with carrier function q.

that corresponds both to the grooves' relative detour the $\alpha(\mathbf{x})$ (see Fig. 2.26) and to the decrease in the grooves' maximum height by the factor $\beta a(\mathbf{x}) \in [0,1]$ (see Fig. 2.27) as compared with the basic grating, where β is a coefficient of the depth of modulation, $\beta \in [0,1]$, $\alpha_1(\mathbf{x}) \in [0,1]$, $\alpha_2(\mathbf{x}) \in [0,1]$, $b(\mathbf{x})$ is a non-modulated component of the phase function that corrects the phase of reconstructed beam.

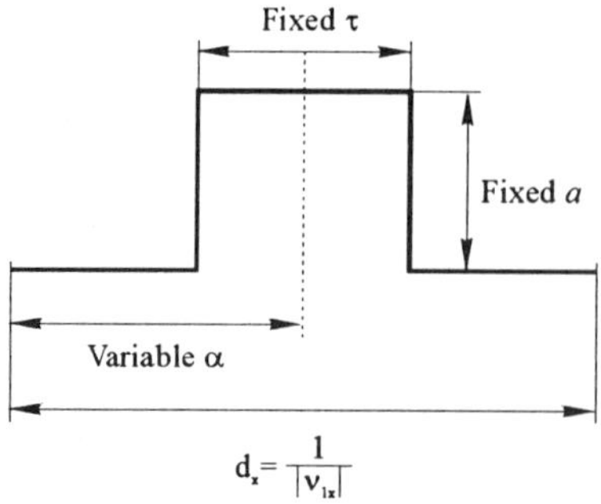


**Figure 2.26.** Modulation of the groove's detour only, for encoding the phase transfer.

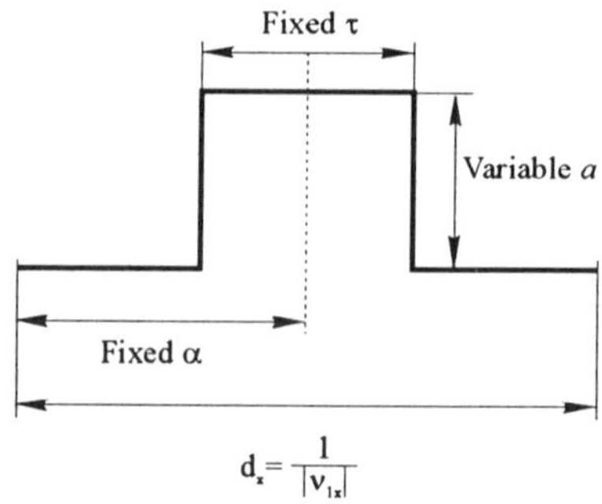


**Figure 2.27.** Modulation of the groove's height only, for encoding the amplitude transfer.

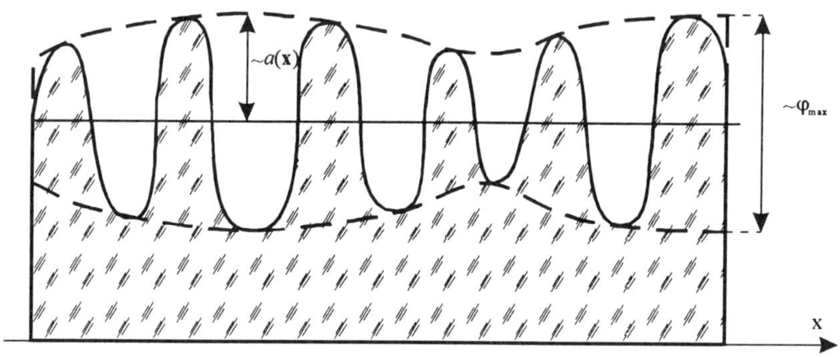


**Figure 2.28.** Diffractive coding with maximum-height modulation. *a(x)*- modulation amplitude, $\varphi_{\max}$- maximum phase delay introduced by modulated phase grating.

Parameters $\tau$ of the $q$ function are also modulated, $\tau = \tau(\mathbf{x})$, if present. Certainly, the functions b(**x**), $a$(**x**), $\alpha$(**x**), $\tau$(**x**) must be slowly-varying as compared with the basic diffraction grating.

The equation of coding (2.35) really describes a non-regular phase diffraction grating (see Fig. 2.28) that becomes a regular one at the special case $b \equiv 0$, $\alpha \equiv 0$ and $a \equiv 1$.

**C. Reconstruction of digital holograms containing complex-to-phase coding**

Such a cross-grating modan represents, in effect, a phase hologram of a modal beam that is described by the complex amplitude $t(\mathbf{x})$ in the modal plane and reference beam $w_0(\mathbf{x})$. Note that the phase hologram can be realized in both transmission (Fig. 2.29) and reflection (Fig. 2.30) variants. Let us show that using suitable selection of the $b(\mathbf{x})$, $\alpha(\mathbf{x})$ and $a(\mathbf{x})$ functions, in Eq.(2.35), one can ensure the required amplitude-phase transfer $T$ in the $\boldsymbol{n}$-th diffraction order ($\mathbf{n} = (n_1, n_2)$ is a 2-D multiplex).

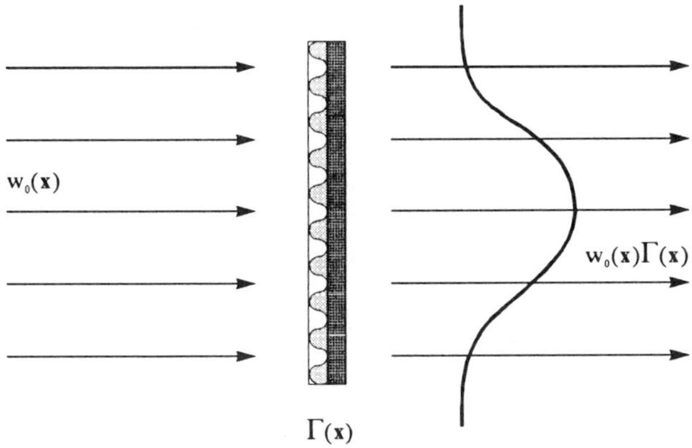


**Figure 2.29.** Transmission-type phase holographic optical element.

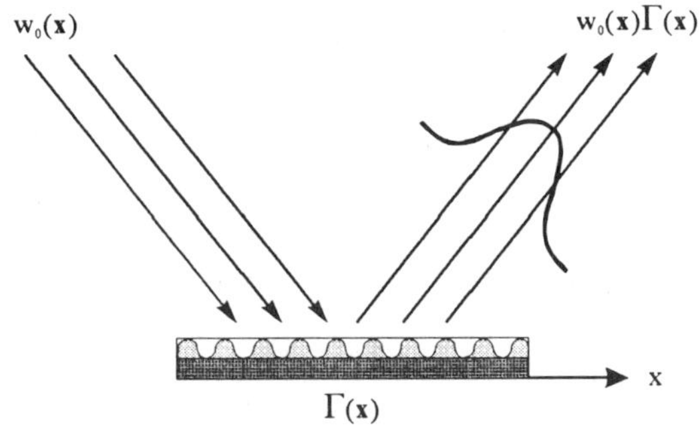


**Figure 2.30.** Reflection-type phase holographic optical element.

Let us take the expansion of the function Γ, Eqs.(2.34) and (2.35), in terms of diffraction orders $\Gamma_{\mathbf{n}}$ of basic diffraction grating:

$$\Gamma(\mathbf{x}) = \sum_{\mathbf{n}} \Gamma_{\mathbf{n}}(\mathbf{x}) \tag{2.36}$$

For this we are forgetting for a while, about coordinate $\mathbf{x}$ in Eq.(2.35) and build the expansion of the periodic function $q$ with period 1 into the 2-D Fourier series

$$\exp\left[i\varphi_{\max} A q(\mathbf{t};\boldsymbol{\tau})\right] = \sum_{\mathbf{n}} B_{\mathbf{n}}(A,\boldsymbol{\tau}) \exp(i2\pi\mathbf{n}\mathbf{t}), \quad A \in [0,1] \tag{2.37}$$

$$B_{\mathbf{n}}(A,\boldsymbol{\tau}) = \int_{0.5}^{0.5} \int_{0.5}^{0.5} \exp[i\varphi_{\max} A q(\mathbf{t};\boldsymbol{\tau}) - i2\pi\mathbf{n}\mathbf{t}]\, d^2\mathbf{t} \tag{2.38}$$

Now we remember about slow dependence of the parameters $\boldsymbol{\tau}$, $a$, $\boldsymbol{\alpha}$ in Eq.(2.35) on coordinates. Owing to Eqs.(2.37), (2.35) and (2.34), the partial transfer function operating in the **n**-th diffraction order is described by equation

$$\Gamma_{\mathbf{n}}(\mathbf{x}) = B_n[\beta a(\mathbf{x};\boldsymbol{\tau}(\mathbf{x}))]\exp\{i[b(\mathbf{x}) + 2\pi\mathbf{n}\boldsymbol{\alpha}(\mathbf{x}) + 2\pi\mathbf{v}_{\mathbf{n}}\mathbf{x}]\}, \quad \mathbf{v}_{\mathbf{n}} = \hat{S}^*\mathbf{n}$$

The amplitude transmittance obtained is non-linear with respect to the value of $a$

$$\frac{B_{\mathbf{n}}\left|\beta a(\mathbf{x};\boldsymbol{\tau}(\mathbf{x}))\right|}{g_{\mathbf{n}}}; \quad g_{\mathbf{n}} = \max_{a\in[0.1]}\left|B_{\mathbf{n}}[\beta a, \boldsymbol{\tau}]\right| \tag{2.39}$$

Provided that the function $\Gamma(\mathbf{x})$ is phase-only, the $\Gamma_{\mathbf{n}}(\mathbf{x})$ component is the amplitude-phase function.

If the following equations are satisfied

$$\left|B_{\mathbf{n}}[\beta a, \boldsymbol{\tau}]\right| = g_{\mathbf{n}}|T| \tag{2.40}$$

$$b(\mathbf{x}) + 2\pi\mathbf{n}\boldsymbol{\alpha}(\mathbf{x}) + \arg B_{\mathbf{n}}[\beta a(\mathbf{x}), \boldsymbol{\tau}(\mathbf{x})] = \varphi_T(\mathbf{x}), \tag{2.41}$$

then the required amplitude-phase transfer $T$ is available the **n**-th operating diffraction order in the form including the constant factor $g_{\mathbf{n}}$ and the carrier frequency $\mathbf{v}_{\mathbf{n}}$:

$$\Gamma_{\mathbf{n}}(\mathbf{x}) = g_{\mathbf{n}}T(\mathbf{x})\exp(i2\pi\mathbf{v}_{\mathbf{n}}\mathbf{x}) \tag{2.42}$$

A number of applications necessitate the development of the amplitude optical elements by means of the linear coding of $T$ complex function into an amplitude form with the cosinusoidal basic diffraction grating with transfer function:

$$\Gamma_a(\mathbf{x}) = T_0 + 2\Delta T|T(\mathbf{x})|\cos[2\pi\boldsymbol{\nu}_1\mathbf{x} + \varphi_T(\mathbf{x})] \tag{2.43}$$

where

$$T_0 = A_0 + \frac{\Delta A}{2}, \quad \Delta T = \beta\alpha\frac{\Delta A}{4},$$

and $[A_0, A_0 + \Delta A] \subset [0,1]$ is the amplitude transmittance range used, $0 \le \beta_a \le 1$. Here an operating diffraction order is the first one with the transfer as

$$\Gamma_{a,1}(\mathbf{x}) = g_a T(\mathbf{x})\exp(2\pi\boldsymbol{\nu}_1\mathbf{x}), \quad g_a = \Delta T$$

The multibeam amplitude modans represent a mathematical superposition of modulated diffraction gratings and can ensure the required separation of the branches in the first diffraction order, but possess low energy efficiency.

There is a special case of coding the real alternating-sign function $T(\mathbf{x})$ into the nonnegative (amplitude only) one. In this case we can simply add the constant to $T(\mathbf{x})$:

$$\Gamma_a(\mathbf{x}) = T_0 + 2\Delta T \cdot T(\mathbf{x}) \quad |T(\mathbf{x})| \le 1 \tag{2.44}$$

to make $\Gamma_a$ nonnegative. In this case operating diffraction order is zero order. However, formula (2.44) can be formally derived from Eq.(2.45) at $\mathbf{v}_1 = 0$. Low energy efficiency of the amplitude spatial filter (2.44) and overlapping the gray background $T_0$ on $T(\mathbf{x})$ are the drawbacks of this simple method.

**D. General approaches for the solution of coding equations**

Let us analyze the equations of the phase coding, i.e., (2.40) and (2.41). It seems expedient to choose the parameter $\varphi_{\max}$ so that the variable $|B_n|$ depends monotone on the variable $a$ on the whole interval $a \in [0,1]$, and $\tau(\mathbf{x}), \mathbf{x} \in G$. Thus, there should not be any zeros of the $|B_n|$ function in the interior points. In this case the function is uniquely related to $|T|$ by the point-to-point monotone nonlinear predistortion, Eq.(2.40).

Equation (2.41) has a variety of solutions because it includes three unknown functions, namely, $b(\mathbf{x})$, $\boldsymbol{\alpha}(\mathbf{x}) = (\alpha_1(\mathbf{x}), \alpha_2(\mathbf{x}))$ and parameters $\tau(\mathbf{x})$. If there is a possibility to realize a hologram whose microrelief height varies continuously it seems natural to modulate only the height of microrelief by changing $a(\mathbf{x})$ without changing the position of the groove's centers. That is, to take $\alpha \equiv 0$, fix $\boldsymbol{\tau}$ and choose

$$b(\mathbf{x}) = \varphi_T(\mathbf{x}) - \arg B_{\mathbf{n}}[\beta a(\mathbf{x}), \tau]$$

In this case the phase $\varphi_T(\mathbf{x})$ is added (with correction) directly to the phase function (2.35) of the holographic optical element, where $a(\mathbf{x})$ is found from Eq.(2.40).

However, if the phase of hologram (including the binary one) can have only a small number of values, it seems reasonable to put $b \equiv 0$ and make use of the detour-phase method, this is that where we introduce the shift $\boldsymbol{\alpha}(\mathbf{x})$ of the grating grooves satisfying the following equation:

$$2\pi \mathbf{n}\boldsymbol{\alpha}(\mathbf{x}) = \varphi_T(\mathbf{x}) - \arg B_{\mathbf{n}}[\beta a(\mathbf{x}), \tau]$$

Use of the combined method suggested here is free of the latter limitations. In this method (Fig. 2.31) the rough phase transfer occurs due to adding $\varphi_T(\mathbf{x})$ to $\varphi(\mathbf{x})$, ($\alpha \equiv 0$). The more accurate coding is implemented by small detours of grooves. From the realization considerations, this method appears to be more simple, however, it fails to operate in the axial (**n**=0) operating diffraction order characterized by the maximum efficiency, and can lead to the overlap of the neighboring grooves on hologram corresponding to the remaining phase according to the general equation (2.41). A combined method is preferable when the number of phase quantization levels is small (ranging from 3 to 8) as in simple photolithography.

Karpeev S.V. in the paper [Beresny, 1991] suggested a binary-phase coding method based on the modulation of the relative groove-width $\tau$ for amplitude-transfer control and groove's detour for phase-transfer control (see Figs. 2.32-2.35). The influence of the relative groove-width $\tau$ on the reconstructed amplitude is given by formula in the third line of Table 2.3.

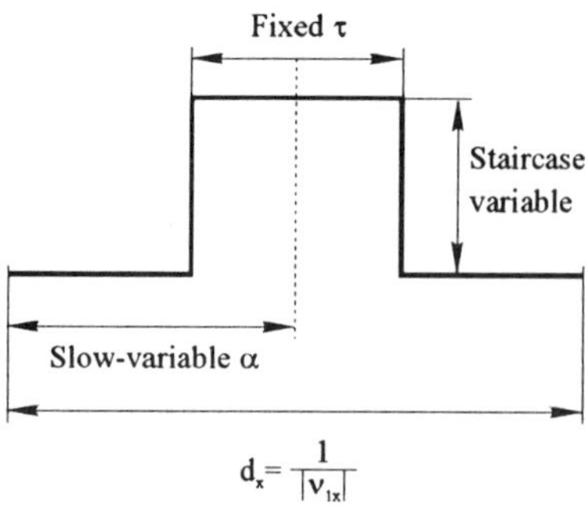


**Figure 2.31.** Combined modulation of the groove's staircase height and small detours.

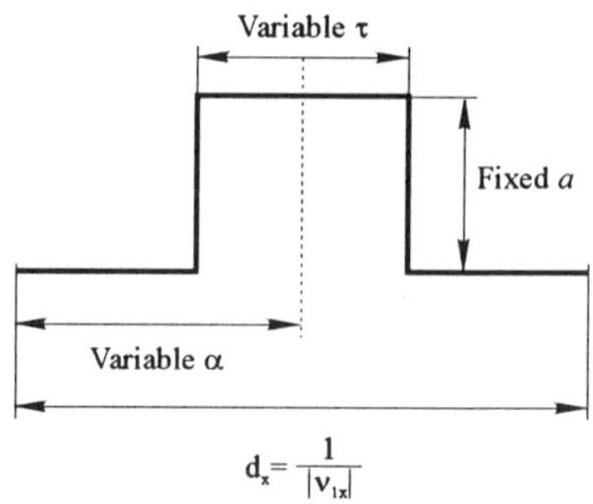


**Figure 2.32.** Modulation of the groove's relative width and detour for encoding the amplitude-and-phase transfer.

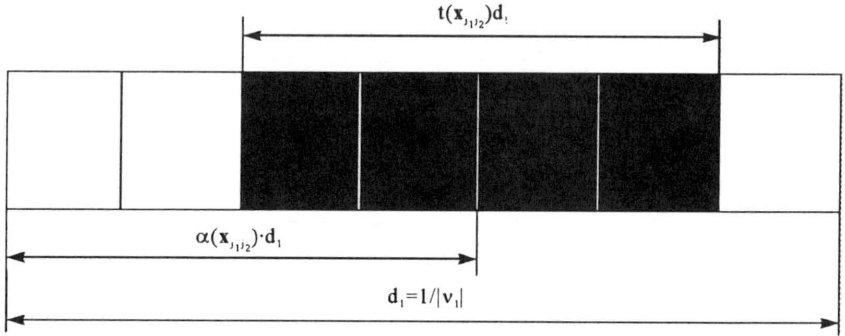


**Figure 2.33.** The principle of binary phase coding of the complex value $T$ and the structure of coding cell. The relative width $\tau$ of the groove is proportional to the $|T|$, the relative detour is proportional to the phase arg$T$. (Modified from Berezny A.E., Karpeev S.V., Uspleniev G.V., Computer generated holographic optical elements produced by photolithography, *Optics and Lasers in Engineering*, 15(5), 331, 1991).

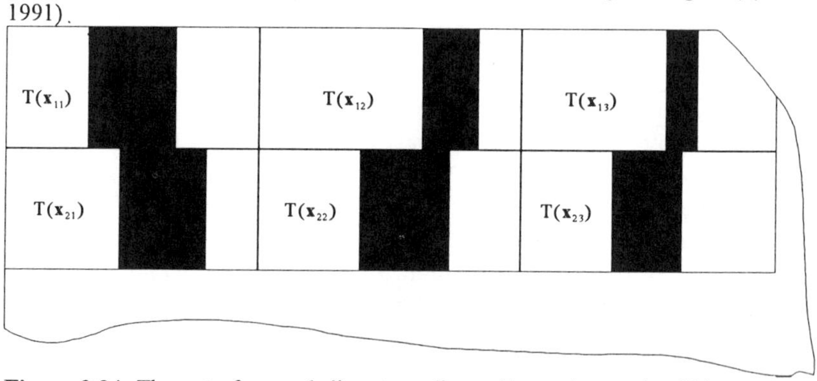


**Figure 2.34.** The set of several discrete coding cells on the mask of binary-phase modan with detour phase and relative groove-width modulation. (Modified from Berezny A.E., Karpeev S.V., Uspleniev G.V., Computer generated holographic optical elements produced by photolithography, *Optics and Lasers in Engineering*, 15(5), 331, 1991).

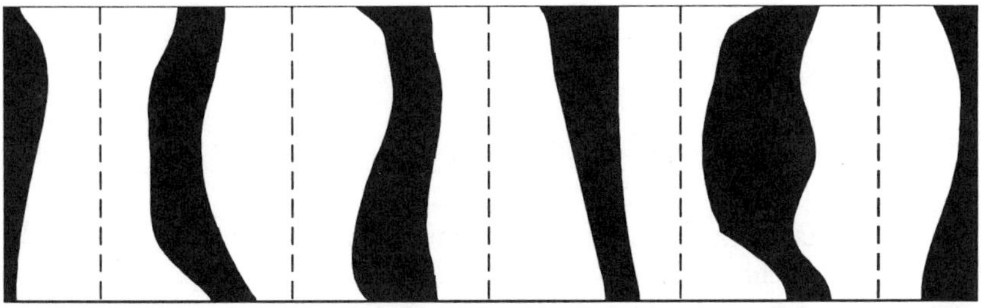

**Figure 2.35.** The fragment of the mask for binary-phase modan with detour phase and relative groove-width modulation. (Modified from Berezny A.E., Karpeev S.V., Uspleniev G.V., Computer generated holographic optical elements produced by photolithography, *Optics and Lasers in Engineering*, 15(5), 331, 1991).

**Table 2.3.**
**Formulae for various types of basic diffraction gratings**

| Basic grating type, $q_0(t)$ | $\varphi_{max}$ | $B_n(A)$ | $g_n$ |
|---|---|---|---|
| Sinusoidal $1/2\cos(2\pi t)$ | ~π | $i^n J_n\left(\frac{\varphi_{max}}{2}A\right)$ | $\max_{t\geq 0}\lvert J_n(t)\rvert$ |
| Rectangular-pulsed with fixed $\begin{cases} 0.5, & l \leq t < 0.5+l \\ -0.5, & 0.5+l < t \leq 1+l \end{cases}$ $l = 0, \pm 1, \pm 2, \ldots$ | π | $\cos(\pi/2\,A),\ n=0$<br>$\frac{2}{\pi n}\sin\left(\frac{\pi}{2}A\right), n = \pm 1, \pm 3, \ldots$<br>$0, n = \pm 2, \pm 4, \ldots$ | 1<br>$2/\pi n$<br>0 |
| Rectangular-pulsed with variable relative width $\begin{cases} 0.5, & l \leq t < \tau+l \\ -0.5, & \tau+l \leq t < 1+l \end{cases}$ $0 \leq \tau \leq 0.5$ | π | $-\frac{2}{\pi}\sin(\pi\tau)\, e^{i\pi\tau}$<br>$(n=1)$ | $\frac{2}{\pi n}$ |
| Sawtooth $(t-0.5-l),\ l \leq t \leq 1+l$ $l = 0, \pm 1, \pm 2, \ldots$ | 2π | $(-1)^n \operatorname{sinc}(A-n)$ | 1 |

**E. Coding with 1-D periodic structures**

As an example, we consider the coding formulae[Golub, 1992a] for 1-D basic grating with

$$q(t_1, t_2) \equiv q_0(t_1 + t_2), \quad q(\hat{S}\mathbf{x}) \equiv q_0(\mathbf{v}_1\mathbf{x}),$$

where $q_0(t)$ is the periodical function of one variable, the period being equal 1. Furthermore, for $\mathbf{n} = (n_1, n_2)$

$$B_{\mathbf{n}}(A) = \delta_{n_1 n_2} \cdot B_{n_1}(A); \quad g_{\mathbf{n}} = g_n,$$

where

$$B_n(A) = \int_{0.5}^{0.5} \exp[i\varphi_{\max} A q_0(t) - i2\pi n t] dt.$$

The calculation formulae for various 1-D diffraction gratings are presented in Table 2.3 and illustrated in Figs. 2.22. Of course, other carrier structures can be used also, for example multiple-pulsed per period.

The nonlinearity of coding is illustrated in Fig. 2.36. Note that a specific case with the cosinusoidal carrier frequency and $n=0,\ \nu_{1x} = \nu_1,\ \nu_{1y} = 0$ gives us the equations equivalent to the Kirk-Jones method [Kirk, 1971; Akaev, 1977]. Another specific case with the rectangular-pulsed carrier for n=0 and

$$\boldsymbol{\nu}_1 = \left( \frac{1}{2\delta}, \frac{1}{2\delta} \right)$$

yields the synthetic coefficients' method [Chu, 1974] or a similar method for the computation of phased antenna gratings that has been developed by [Kashin, 1970]. The coding based on the sawtooth carrier grating allows of operating in the first diffraction order with the maximum possible efficiency of coding ($g_n^2 = 1$) that features only the zero order usually.

**F. Light field reconstructed from modan**

Let the holographically synthesized modan with the phase-only transfer be illuminated with the launching beam. The total field reconstructed just behind the plane is equal to

$$w(\mathbf{x}) = w_0(\mathbf{x})\Gamma(\mathbf{x})$$

and, allowing for Eqs.(2.36), (2.42) and (2.32), the field in the n-th working order may similarly be written in the form

$$W_{\mathbf{n}}(\mathbf{x}) = w_0(\mathbf{x})\Gamma_{\mathbf{n}}(\mathbf{x}) = \sqrt{E_0} g_{\mathbf{n}} c t(\mathbf{x}) \exp(i2\pi \boldsymbol{\nu}_{\mathbf{n}} \mathbf{x}) \qquad (2.45)$$

This field differs from $t(\mathbf{x})$ only by a constant factor proportional to $g_{\mathbf{n}} \cdot c$ . An additional carrier frequency $\boldsymbol{\nu}_{\mathbf{n}}$ describes the tilt of the reconstructed beam of the **n**-th diffraction order.

The additional carrier spatial frequency $\nu_{\mathbf{n}}$, which is more than $\nu^{(\mathbf{k})}$, "turns" all the partial beams in branches as a unity without changing their interaction. If we use the tandem components, with one of them generating $|T|$ and with the other generating the phase $\arg T$, we can assume in Eq.(2.45) that $g_{\mathbf{n}} \cdot c = 1$.

Equations (2.45) and (2.2) give us relationships

$$W_{\mathbf{n}}(\mathbf{x}) = \sum_{\mathbf{k} \in K} W_{\mathbf{n}}^{(\mathbf{k})}(\mathbf{x}), \quad k \in \mathrm{K} \qquad (2.46)$$

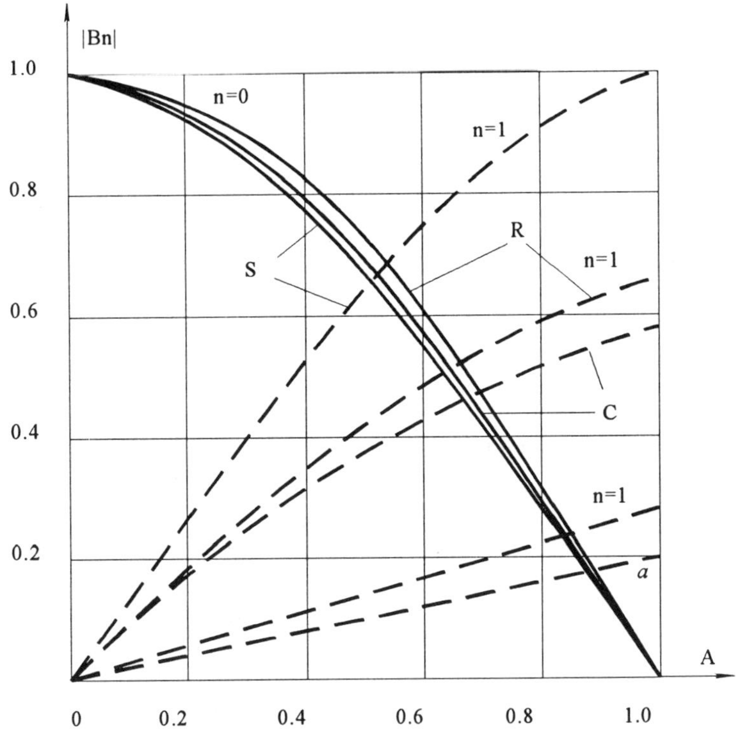


**Figure 2.36.** Nonlinearity of amplitude control in complex-to-phase coding. *n* is the number of the operating diffraction order; *S,C,R* letters mark the cases of phase coding with the following groove shapes of basic gratings; *C*-cosinusoidal; *S*-sawtooth; *R*-rectangular-pulsed; *a*-letter marks (for comparison) the case of amplitude cosinusoidal diffraction grating with linear but inefficient amplitude control.

$$W_{\mathbf{n}}^{(\mathbf{k})}(\mathbf{x}) = \sqrt{E_0} g_{\mathbf{n}} c \mu^{(\mathbf{k})} \frac{\xi^{(\mathbf{k})}(\mathbf{x})}{\left|\Xi^{(\mathbf{k})}\right|} \exp[i2\pi \mathbf{v}_{\mathbf{n}}^{(\mathbf{k})}\mathbf{x}] \tag{2.47}$$

$$\mathbf{v}_{\mathbf{n}}^{(\mathbf{k})} = \mathbf{v}_{\mathbf{n}} + \mathbf{v}^{(\mathbf{k})} \tag{2.48}$$

Equation (2.47) makes it possible to conclude that each branch at the modan's output with the n-th working diffraction order has the definite direction defined by the vector of spatial frequencies and by the unit 3-D vector

$$\mathbf{N}_{\mathbf{n}}^{(\mathbf{k})} = \left( \lambda \mathbf{v}_{\mathbf{n}}^{(\mathbf{k})}, \sqrt{1-\lambda^2[\mathbf{v}_{\mathbf{n}}^{(\mathbf{k})}]^2} \right).$$

The set of vectors $\mathbf{N}_{\mathbf{n}}^{(\mathbf{k})}$, $\mathbf{k} \in K$ is turned as a unity with respect to the required vectors $\mathbf{N}^{(\mathbf{k})}$ of the branch directions.

**G. Estimation of achievable number of modan's branches**

The mixture of different spatial frequencies and the auxiliary frequency that are present in the equations of the phase modan discussed in section 2 requires a thorough examination of conditions of the angular separation for the different diffraction orders of modan. The procedures of the phase sampling and quantization employed in digital holography complicate the situation, too.

To solve the problem of the angular separation in terms of estimation, let us consider the angular spectrum specified as the two-dimensional Fourier transform of the complex amplitude immediately behind the modan plane

$$S(\mathbf{v}) = \int W_{\mathbf{n}}(\mathbf{x}) \exp(-i2\pi \mathbf{v}\mathbf{x})\, d^2\mathbf{x}. \tag{2.49}$$

Let us consider here only the most important case of zero-order operating diffractive order. Thus, we have $\mathbf{n}=0$, $\boldsymbol{\nu}_{\mathbf{n}}=0$, $\boldsymbol{\nu}_{\mathbf{n}}^{(\mathbf{k})}=\boldsymbol{\nu}^{(\mathbf{k})}$

From Eqs.(2.46) and (2.49) we obtain

$$S(\mathbf{v}) = \sum_{\mathbf{k}\in K} S^{(\mathbf{k})}(\mathbf{v}-\mathbf{v}^{(\mathbf{k})}). \tag{2.50}$$

Let the angular spectrum of the **k**-th modal beam have the width of $b^{(\mathbf{k})}$. We introduce the parameters

$$b = \max_{\mathbf{k}\in K} b^{(\mathbf{k})}$$

Figure 2.37 depicts the position of the angular spectra for the separate modal beams. The conditions of separation (2.4) can be written here in the form

$$\left|\mathbf{v}^{(\mathbf{k})} - \mathbf{v}^{(\mathbf{k}')}\right| \geq b.$$

One should also bear in mind that there is a limiting spatial frequency $\nu_{\max}$ of the recording medium. Then the achievable of the modan branches, $L_0 \max$ is equal to the number of the circles with the diameter $b$ that fall into the circle with the radius $\nu_{\max}$ (see Fig. 2.29). Note that plus-and-minus-first diffraction orders in thin holograms cause the points $\mathbf{v}^{(\mathbf{k})}$ to be placed only on one (i.e., the top) half-plane. Then we obtain the following estimation

$$L_0 \max \leq \frac{\pi \nu_{\max}^2}{2b^2}. \tag{2.51}$$

In a number of cases the requirement of equal power in branches necessitates satisfying the condition $\left|\mathbf{v}^{(\mathbf{k})}\right| = \nu_0 = const$, that is, all the values $\mathbf{v}^{(\mathbf{k})}$ are placed on the semi-circumference of radius $\nu$. Here the maximum number of the branches is reduced to

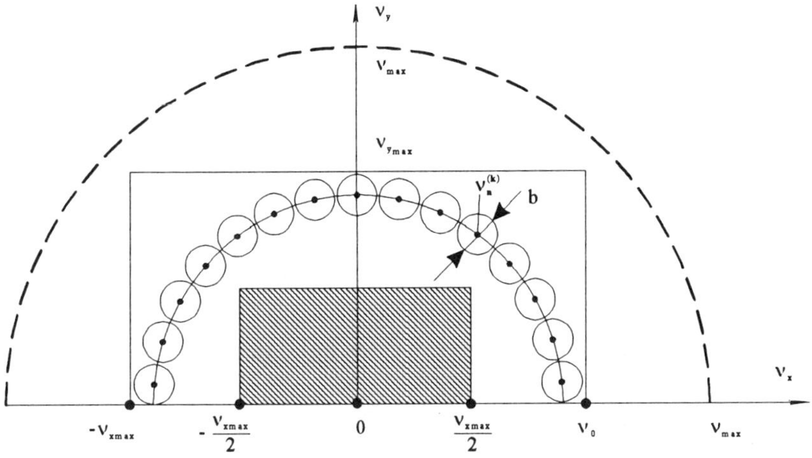


**Figure 2.37.** Fourier plane, illustrating the angular separation of the branches for multibeam modan. $(\nu_x, \nu_y)$-are the spatial frequencies, $\nu_{max}$-maximum spatial frequency of the recording medium, $\nu_{xmax}$, $\nu_{ymax}$ are the maximum spatial frequencies of the modan's branches, $\boldsymbol{\nu}_{\mathbf{n}}^{(\mathbf{k})}$ -the carrier spatial frequencies of the **k**-th branch. (Modified from Adzhalov V.I., Golub M.A., Karpeev S.V., Sisakyan I.N., Soifer V.A., Multi-channel computer optics components matched to mode groups, *Soviet Journal of Quantum Electronics*, 20(2), 136, 1990).

$$L_0' \max = \frac{\pi \nu_0}{b} \leq \frac{\pi \nu_{max}}{b}. \tag{2.52}$$

Actually, the size $\tilde{\sigma}$ of the angular spectrum of the principal mode is related to σ by the well-known equation $\pi\sigma\tilde{\sigma} = 1$. Let us choose a limitation of the spectrum size $b$ on the same intensity level as that obtained from the size $D$ in a modan plane, therefore $D/\sigma = b/\tilde{\sigma}$.

Therefrom we have

$$b = \frac{D}{\pi\sigma^2} = \frac{4L_m}{\pi D},$$

where $L_m = (D/2\sigma)^2$ is a dimensionless parameter that determines the maximum order of the mode that falls into the transverse size D of the modan.

For an optical system with the quadratic graded-index fiber we have $D/2\sigma = a/\sigma_0$ and $L_m = (a/\sigma_0)^2$, where $a$ and $\sigma_0$ are the radii of the core and of the fiber's principal mode, respectively. The value $L_m$ is in agreement with the parameter of normalized frequency for the graded-index fiber that has been described by [Unger, 1977]. Estimation (2.51) enables us to analyze what physical limitations exist when we perform multiplexed recording of the spatial filters on the substrate having a limited resolution $\nu_{max}$. For the typical values $\nu_{max}$=200 lines/mm, $L_m$=50, $D$=6.4 mm we obtain a number of the simultaneously analyzed transverse modes $L_0$ max=600. Extra limitations $\nu_{max} \leq \frac{1}{\delta}$ on the spatial frequency $\nu_{max}$ arise in computer-aided synthesis of modans where δ is the resolution of the mask recording device. This extra limitation results in the essential decrease of $L_{0_{max}}$. For $\delta_1$=25μm, $\nu_{max}$=40 lines/mm, $L_m$=50, $D$=6.4 mm we have $L_{0_{max}}$=25, that is more than by an order greater than the value caused by physical limitations.

If $\boldsymbol{\nu}^{(\mathbf{k})}$ is placed in the semi-circumference, according to Eq.(2.52) we obtain $L'_{0_{max}}$=12, for $\nu_0$=40 lines/mm, $L_m$=50, and $D$=6.4 mm.

# Chapter 3

# COHERENT MODES SELECTION

## 3.1. OPTICAL STAGES FOR MODE SELECTION WITH THE AID OF MODANS

Dealing with laser cavities, lens-like media, and fiber waveguides, we can find the beams with various power distribution between modes [Pressley, 1971]. At the same time there are practical problems where the selective operation with a single mode or a certain modal group is needed. For example, fiberoptic communication systems encounter a problem of measuring or/and correcting for the differential attenuation of modes as well as of their differential modal delays, which result in the blurring of the pulse [Kapany, 1967; Unger, 1977]. Formally speaking, we deal with measuring and correcting of the amplitude and phase of the coefficients of light beam expansion in mode series. In other words we must think about such operations as modal analysis and filtration in modal representation. Another problem arises upon forming a time-dependent light beam for a fiberoptic communication line involving the modal multiplexing of channels. In this case, one must be able to perform independent temporal-variations of the modal powers and to realize a modal superposition keeping pace of data transmission.

Let us state the three basic types of the problems in the mode selection:
- first, the modal analysis, that is, the separation of a multimodal light beam into separate modal components and the measurement of the power distribution between modes as well as of intermodal phase shifts (see Fig. 3.1);
- second, the mode formation (synthesis), that implies excitation of standards of the required modes or multimodal beams with a required power and phase distribution between modes (see Fig. 3.2);
- third, the modal filtration based on modal transforms that changes the modal composition in both the power and the phases (see Fig. 3.3).

Special role in light-beam transformations is played by the cosinusoidal (trigonometric) modes. For this case the mode analysis is performed using a common Fourier stage (see Fig. 3.4), the mode formation is carried out by the similar inverse Fourier stage (Fig. 3.5) and the mode filtration is performed using a two-stage correlator with the help of a complex spatial filter in the Fourier-plane (Fig. 3.6).

To make the evidence of the said above let us represent the expansion in modes, Eqs.(1.8) and (1.9), using a double trigonometric Fourier series

$$w(\mathbf{x}) = (2a\cdot 2b)^{-\frac{1}{2}} \sum_{p,l=-\infty}^{\infty} w_{pl} \exp(i2\pi \nu^{(pl)}\mathbf{x})$$

and an integral equation for the computation of its coefficients

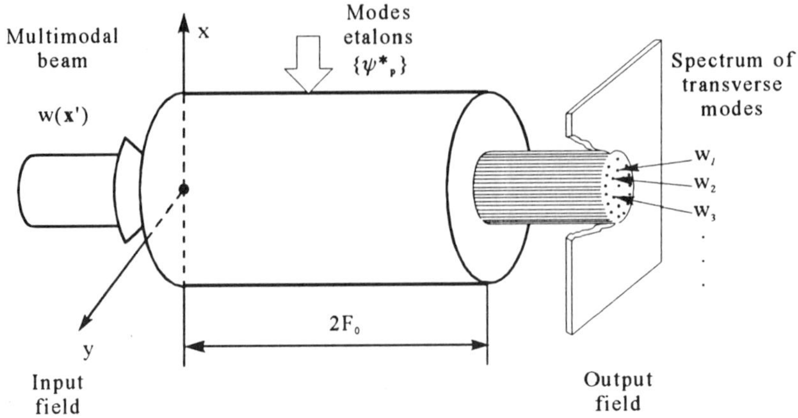


**Figure 3.1.** Modal analysis in general .

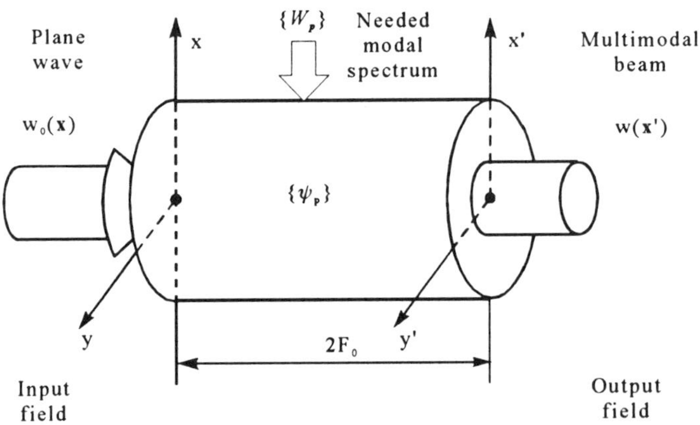


**Figure 3.2.** Modal formation (synthesis) in general .

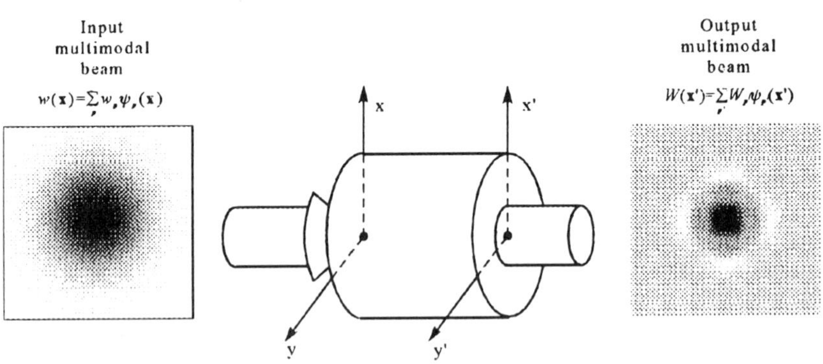


**Figure 3.3.** Modal filtration in general .

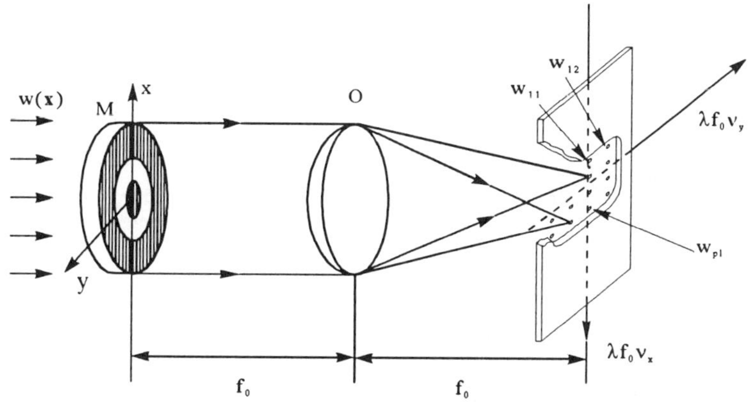


**Figure 3.4.** Implementation of modal analysis and direct modal transform on the base of optical Fourier stage. *w(**x**)* - input multimodal beam, *O* - objective, $M$ - modan, $w_{pl}$ - output coefficients for mode number *(p,l)*, $f_0$ - focal length of *0* objective.

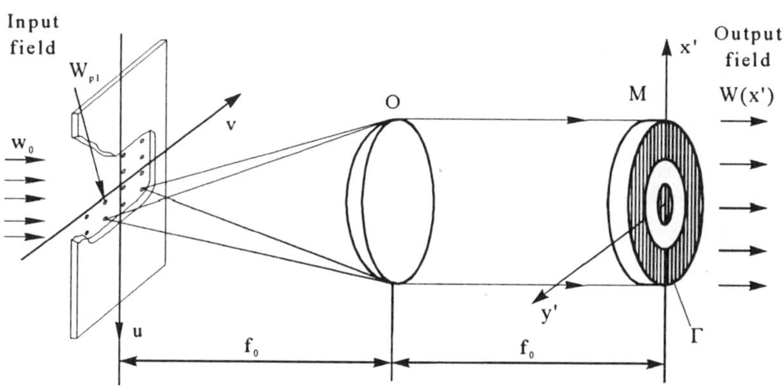


**Figure 3.5.** Implementation of inverse modal transform on the basis of optical Fourier stage. $w_0$ - illuminating beam, *O* - objective, $M_2$ - modan, $w_{pl}$ - input coefficients for mode number *(p,l)*, *W(**x**')* - output multimodal beam, $f_0$ - focal length of *0*.

$$w_{pl} = (2a \cdot 2b)^{-\frac{1}{2}} \int_{-a}^{a} \int_{-b}^{b} w(\mathbf{x}) \exp(-i2\pi \boldsymbol{\nu}^{(pl)}\mathbf{x}) d^2\mathbf{x} \tag{3.1}$$

in a rectangular region $G$ that measures $2a \times 2b$, where $\boldsymbol{\nu}^{(pl)}$ is a vector of multiple spatial frequencies specified by Eqs.(1.113) and (1.114). At the same time, if there is a screen with the rectangular aperture $G$ at the input plane, the Fourier-stage for the direct transform is specified by equation

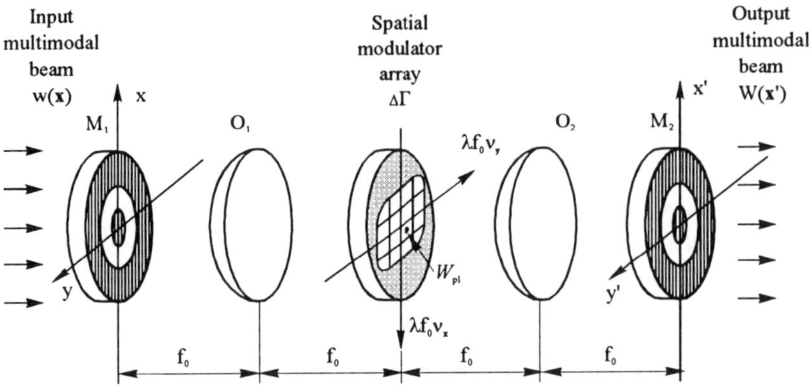


**Figure 3.6.** Implementation of modal filtration on the base of optical correlator. *w(**x**)* - input multimodal beam, $O_1, O_2$ - objectives, $M_1, M_2$ - modans, ΔΓ - array of spatial modulators, *W(**x**')* - output multimodal beam.

$$W(\boldsymbol{\nu}) = \int_{-a}^{a}\int_{-b}^{b} w(\mathbf{x})\exp(-i2\pi\boldsymbol{\nu}\mathbf{x})\, d^2\mathbf{x} \tag{3.2}$$

Comparison between Eqs.(3.1) and (3.2) gives us the solution of the problem of mode analysis in the form

$$w_{pl} = (2a\cdot 2b)^{-\frac{1}{2}} W(\boldsymbol{\nu}^{(pl)}).$$

This means that the modal coefficients $\{w_{pl}\}$ are proportional to the values of the Fourier-image amplitude on the discrete grid of the equidistant points, furthermore, the mode powers are equal to the pixels of intensity in the same points (Fig. 3.4).

The problem of the modal formation turns out to be resolvable if we consider a double Fourier-stage input (Fig. 3.6) with a grid of pinholes in points $\boldsymbol{\nu}^{(pl)}$ of frequency plane cutting out the pixels $W(\boldsymbol{\nu}^{(pl)})$ and thus forming the complex amplitude

$$W(\boldsymbol{\nu}) = \sum_{p,l} W(\boldsymbol{\nu}^{(pl)})\delta(\boldsymbol{\nu}-\boldsymbol{\nu}^{(pl)}).$$

Here $\delta(\cdot)$ is 2-D delta-function

At the output plane of second Fourier-stage we then obtain a field

$$W(\mathbf{x}') = (2a\cdot 2b)^{-\frac{1}{2}} \sum_{p,l} w_{pl} \exp(i2\pi\boldsymbol{\nu}^{(pl)}\mathbf{x}'),$$

that agrees, within a constant factor, with the modal beam $w(\mathbf{x})$.

Thus, for sinusoidal modes there is available optical stage for the modal analysis and formation that represents a common-used Fourier-stage with supplemental sampling of its Fourier-plane.

As V.A.Soifer and I.N.Sisakyan have mentioned [Sisakyan, 1990], a reasonable question arises here: whether the Fourier transform is unique or there are special optical stages for doing other optical modal transforms?

In fact, the Fourier-stage contains the standard of trigonometric mode in the form of the field of tilted plane wave. Since the integral of plane waves taken over the polar angle, yields the Bessel function [Papoulis, 1968], the Fourier stage should as well contain the standards of the Bessel modes of zero order ($l$=0), see Eq.(1.111). It has been proposed in [Bereznyi, 1984] that with the aim of developing the standards of the Bessel modes of any order, Eq.(1.111), optical stages and spatial filters should be used.

The development of optical stages capable of selectively affecting the modes of other types depends on the availability of a physical standard of a single mode. The modans used in optical systems (see Figs. 3.4 through 3.6) offer a possibility of solving the problem of modal analysis, formation and filtering. We discuss below the modan operation in optical spatial filtering systems. In terms of hologram's theory, the image reconstruction takes place on the basis of spatially modulated illuminating beams.

## 3.2. LIGHT BEAM FORMATION WITH REQUIRED COMPOSITION OF MODES

The formation of beams with preset modal composition is a matter of topical interest as applied to the tasks of launching the radiation into optical fibers. Selective excitation of some non-linear media and the testing of quantum amplifiers also require such beams.

The beam with the preset modal composition is needed for the precise focusing of laser radiation to record the information on an optical disk as well as to form the beams with the minor divergence in the optical antennas used as laser beams [Golub, 1989a]. The formation of the beam with the Gaussian distribution that is the zero-order mode, is carried out by placing a diaphragm inside a laser cavity [Lissak, 1990]. To form the beams containing modes of other orders, it is proposed to use computer-generated optical elements - modans [Golub, 1982]. Below we discuss several basic schemes of application of modans to fiber optics.

A multimodal light field comprising the required modes arises directly behind the modan plane if the modan computed using Eqs.(2.32), (2.2), (2.6) is illuminated by the beam $w_0$. To achieve matching in scale between the field formed and an optical fiber waveguide, one can use an extra Fourier-stage (see Fig. 3.7). We can employ several available branches of the modan with the angular separation to perform switching of the exciting beam with the help of deflector.

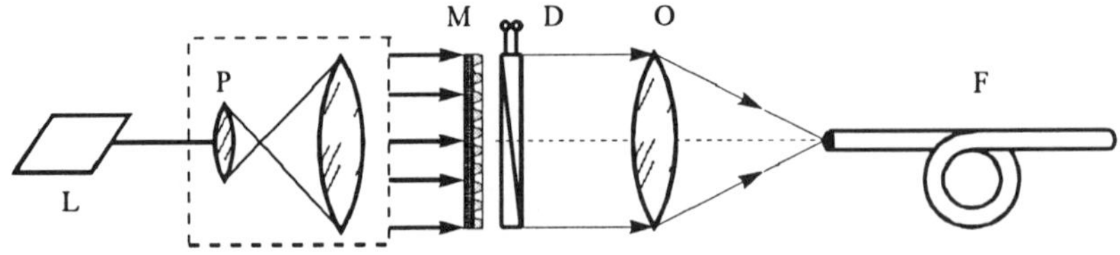


**Figure 3.7.** Mode-selective excitation of a fiber by standard multimodal beam with switching. $L$ - laser, $P$ - pinhole collimator block, $M$ - modan, $D$ - deflector, $O$ - objective, $F$ - fiber. (Modified from Adzhalov V.I., Golub M.A., Karpeev S.V., Sisakyan I.N., Soifer V.A., Multi-channel computer optics components matched to mode groups, *Soviet Journal of Quantum Electronics*, 20(2), 136, 1990).

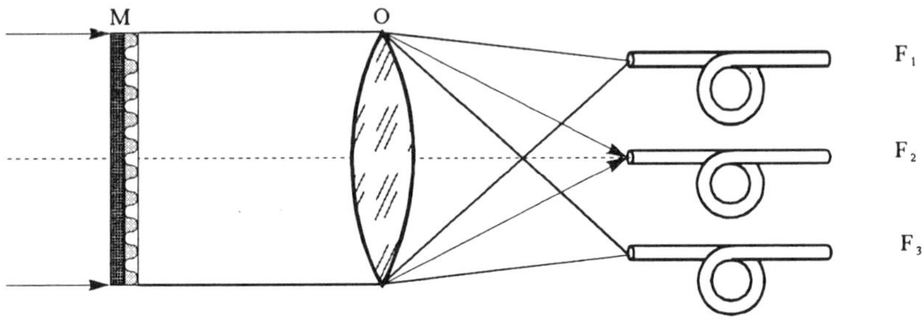


**Figure 3.8.** Parallel excitation of fibers by multibeam modan. $M$ - modan, $O$ - objective, $F_1, F_2, F_3$ - fibers. (Modified from Adzhalov V.I., Golub M.A., Karpeev S.V., Sisakyan I.N., Soifer V.A., Multi-channel computer optics components matched to mode groups, *Soviet Journal of Quantum Electronics*, 20(2), 136, 1990).

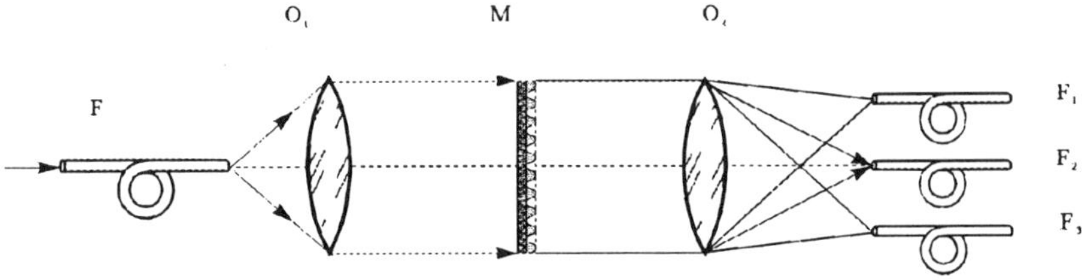


**Figure 3.9.** Coupling of fiber $F$ with several fibers $F_1, F_2, F_3$. Mode selection is performed by modan $M$. $O_1, O_2$ - objectives. (Modified from Adzhalov V.I., Golub M.A., Karpeev S.V., Sisakyan I.N., Soifer V.A., Multi-channel computer optics components matched to mode groups, *Soviet Journal of Quantum Electronics*, 20(2), 136, 1990).

A multibeam modan makes it possible to realize the selective and parallel excitation of several fiber waveguides as shown in Fig. 3.8. In the other application the modan splits the radiation of a single fiber into several fibers and simultaneously performs the mode selection (see Fig. 3.9). Besides, a modan allows of mixing the radiation from several fibers (Fig. 3.10) to be done.

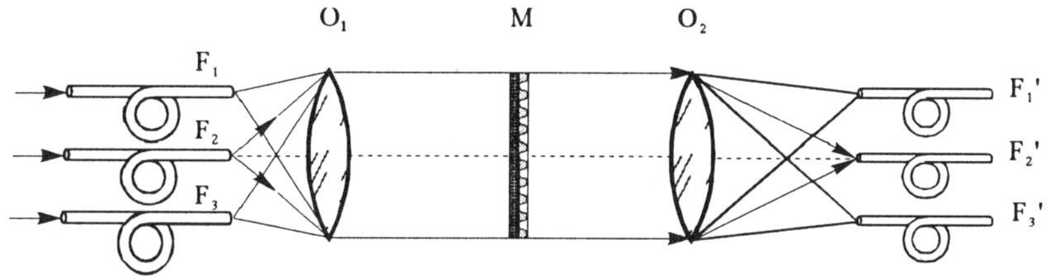


**Figure 3.10.** Inter-fiber mixing of modes. $M$ - modan, $O_1, O_2$ - objectives, $F_1, F_2, F_3$ - input fibers, $F_1', F_2', F_3'$ - output fibers. (Modified from Adzhalov V.I., Golub M.A., Karpeev S.V., Sisakyan I.N., Soifer V.A., Multi-channel computer optics components matched to mode groups, *Soviet Journal of Quantum Electronics*, 20(2), 136, 1990).

A modal formation poses the problem of either excitation of some modes (a single mode, in particular) Fig. 3.7, or synthesis of the full field from a set of the coefficients of its expansion in modes (an inverse modal transform, Figs. 3.2 and 3.5).

## 3.3. THE ANALYSIS OF MODAL COMPOSITION IN COHERENT LIGHT BEAMS

Measurements of the modal structure of a beam that emerges from a laser, turn out to be essential in such laser applications as holography, and holographic interferometry.

Such problems as testing of fiberoptic communication lines and devices, diagnosis of operating conditions and serviceability of high-power lasers as well as studies of the mechanism of operation of newly developed lasers, also necessitates the measurements related to separate modes. The class of modal measurements includes the following measurements: radiation power distribution between modes, differential delays, differential modal attenuation, modal coupling coefficients. Each of the measurements listed above calls for the selectively measuring the power of separate modes $\{\psi_p\}$ or, in some cases, of intermodal phase shifts. That can be treated as an analysis of the modal composition of the field ("modal analysis" for short) propagating in a fiber.

The modal analysis (Figs. 3.1 and 3.4) is capable of solving the task of studying single or several only modal coefficients, Eq.(1.9), of the field under study. Furthermore measuring the total modal composition can be performed. The first approach in measurements is not complete in terms of the beam field, although it can be complete in terms of some other physical parameter, for example, the value of the refractive index profile in a graded-index fiber or the parameter of inter-adjustment of the cavity mirrors.

The second approach can be called a straightforward modal transform. It converts the field complex amplitude in the beam cross-section to a complete set of modal coefficients being ordered, spatially separated and imaged as pixels of the complex amplitude in the corresponding points of the output plane. In such a case a set of measured modal coefficients must provide the high accuracy of the field approximation by the corresponding fraction of series Eq.(1.8), see section 1.1.

The methods for measuring the modal composition of coherent beams based on the digital processing of the field diffraction patterns derived from the fiber output end, require the complicated field intensity scanners to be applied and the digital integral transforms to be performed, and so they are unsuitable for making high-speed analysis of modes.

At the same time, the regular diffraction gratings are known to be used in various spectral devices to decompose the light beams in "longitudinal" modes or wavelengths with only optical processing. A diffraction grating coupled with a lens makes it possible to visualize the power distribution between longitudinal modes in the form of the intensity distribution in the rear focal plane.

The development of the similar spectral device capable of decomposing the light beam in transverse modes is of great interest. First of all, such devices are needed to perform the measurements and investigation of multimodal laser beams propagating in a free space or in a graded-index medium. In the present work it has been proposed to develop "spectral-modal" devices - modans, in the form of curvilinear (non-regular) phase diffraction gratings with continuously varying height of grooves allowing for the spatial modal modulation to be performed.

Assume that $\{\psi_{\mathbf{p}}(\mathbf{x})\}$ are the complex amplitudes of the modes of fiber under study and they have the orthogonality property (1.3) with $\rho_0(\mathbf{x}) \equiv 1$, where $\mathbf{x}=(x,y)$ are the two-dimensional Cartesian coordinates in the beam cross-section and $G$ is a region of the beam working section or light aperture in other words.

We propose to perform in two steps the measurements of the modal composition $\{w_{\mathbf{p}}\}$ of the field, that have the complex amplitude as

$$w(\mathbf{x}) = \sum_{\mathbf{p}} w_{\mathbf{p}} \psi_{\mathbf{p}}(\mathbf{x}), \quad \mathbf{x} \in G, \tag{3.3}$$

The first stage is an optical selection of modes using their orthogonality property. With this, the power distribution between modes is transformed into the light intensity distribution in a focal plane. On the second stage the power of each mode $\sim \left| w_{\mathbf{p}} \right|^2$ is measured using the common photo-receiver with the point aperture.

Let us demonstrate that if we place a spatial filter $M_1$ with the mathematical function of the complex transfer $\psi_{\mathbf{p}}^*(\mathbf{x})$ at the Fourier stage input plane and direct the field $w(\mathbf{x})$ to the input, then the coefficient $w_{\mathbf{p}}$ of a single mode will be formed in the center of the output plane (Fig. 3.4). Indeed, immediately behind the plane $(x,y)$ we obtain the field

$$w(\mathbf{x})\psi_{\mathbf{p}}^*(\mathbf{x}) = \sum_{\mathbf{p}'} w_{\mathbf{p}'}\psi_{\mathbf{p}'}(\mathbf{x})\psi_{\mathbf{p}}^*(\mathbf{x}) \tag{3.4}$$

When taken in the center $\boldsymbol{\nu}=0$ of the output plane of spatial frequencies $\boldsymbol{\nu} = (\nu_x, \nu_y)$, the Fourier transform yields an operation of the complex amplitude integration. By virtue of the orthogonality property, the modes' interaction occurs in such a way that on integrating the complex amplitude (3.4) in the point $\boldsymbol{\nu}=0$ of the Fourier plane, the backgrounds of modes numbered $\mathbf{p}' \neq \mathbf{p}$ cancel each other, and we obtain the field values equal to the coefficient of an expansion

$$w_{\mathbf{p}} = \int w(\mathbf{x})\psi_{\mathbf{p}}^*(\mathbf{x})d^2\mathbf{x} \tag{3.5}$$

The power $\left| w_{\mathbf{p}} \right|^2$ of the mode $\psi_{\mathbf{p}}$ is then measured as an intensity using a conventional photo-receiver with the point aperture. Successive mechanical replacement of the filters $\psi_{\mathbf{p}}^*$ makes it possible to measure the whole modal composition.

To impart a reality to this schematic consideration, we propose the multibeam modan to be used in a modal analyzer of Fig 3.4. Multibeam modan convents available illuminating beam into several titled partial beams called branches. Each partial beam can be single-modal or multimodal containing some desired modal group and in practical set-ups Figs 3.8-3.10.

Figs. (3.11) and (3.12) illustrate the optical set-ups aimed at analyzing the fiber modal composition using the refractive (Fig. 3.11) or reflecting (Fig. 3.12) multibeam modan.

From the formal point of view, the multibeam modan must ensure a parallel (simultaneous) formation of the coefficients $w_{\mathbf{p}}$ in the form of the complex amplitude in a set of the points of focal region. To acquire generality it is required that each branch should comprise not simply a single coefficient $w_{\mathbf{p}}$ but the coefficient's linear combination with the numbers $\mathbf{p}$ from the set $L^{(\mathbf{k})}$ and the complex weight coefficients $\{\xi_{\mathbf{p}}^{(\mathbf{k})}\}$.

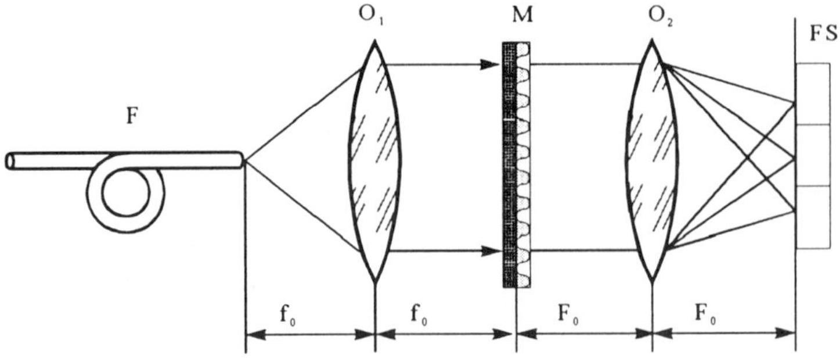


**Figure 3.11.** Analyzer of modal power distribution in fiber with the aid of multibeam transparent modan. $F$ - fiber, $M$ - modan, $O_1, O_2$ - objectives, $FS$ - photoreceiver array. $f_0, F_0$ -focal lengths of $O_1, O_2$ respectively. (Modified from Adzhalov V.I., Golub M.A., Karpeev S.V., Sisakyan I.N., Soifer V.A., Multi-channel computer optics components matched to mode groups, *Soviet Journal of Quantum Electronics*, 20(2), 136, 1990).

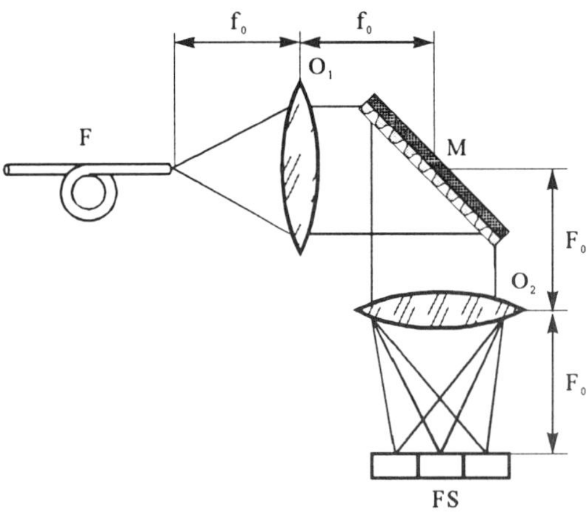


**Figure 3.12.** Analyzer of modal power distribution in fiber with the aid of multibeam reflective modan. $F$ - fiber, $M$ - modan; $O_1, O_2$ - objectives; $FS$ - photoreceiver array; $f_0, F_0$ -focal lengths of $O_1, O_2$ objectives respectively. (Modified from Golub M.A., Sisakyan I.N., Soifer V.A., Uvarov G.V., Optical components for analysis and formation of the transverse mode composition, *Soviet Journal of Quantum Electronics*, 19(4), 543, 1989).

In order to realize the optical computations described above, we place in the scheme in Fig. 3.4 a multibeam modan $M_1$ with the transfer $\Gamma^*(\mathbf{x})$, where the function $\Gamma(\mathbf{x})$ is specified by Eqs.(2.34), (2.35), (2.32), (2.2), and (2.6). In that case the illuminating beam is assumed to be plane, $w_o$=const, and the power distribution in $L$ branches is uniform

$$\kappa^{(\mathbf{k})}=\frac{1}{L},\quad \alpha^{(\mathbf{k})}=0,\quad \mu^{(\mathbf{k})}=\frac{1}{\sqrt{L}} \tag{3.6}$$

If we place a modal analyzer into the Fourier optical set-up (Fig. 3.4), the modan performs a parallel formation of a set of modal beams, which are angular separated near the plane $M_1$ and separated in coordinates in the Fourier-conjugate plane $(\nu_x,\nu_y)$. Let the beam - separation condition (2.4) be satisfied. In that case in the $\lambda f_0 \boldsymbol{\nu}_{\mathbf{n}}^{(\mathbf{k})}$ points of Fourier plane corresponding to the spatial frequencies $\boldsymbol{\nu}_{\mathbf{n}}^{(\mathbf{k})}$, Eq.(2.48), we can find, owing to Eqs.(3.2), (2.46) and (2.47), the complex amplitudes proportional to

$$w^{(\mathbf{k})}=\int_G w(\mathbf{x})W_{\mathbf{n}}^{*}(\mathbf{x})\exp\left[-i2\pi\boldsymbol{\nu}_{\mathbf{n}}^{(\mathbf{k})}\right]d^2\mathbf{x}=\sqrt{\frac{E_o\varepsilon_{\mathbf{n}}}{L}}\left(w,\frac{\xi^{(\mathbf{k})}}{\left|\Xi^{(\mathbf{k})}\right|}\right),\ \varepsilon_{\mathbf{n}}=g_{\mathbf{n}}^2c^2 \tag{3.7}$$

It will be shown in chapter 4 that $\varepsilon_{\mathbf{n}}$ value is really the energy efficiency of modan.

Subject to Eqs.(2.47) and (2.6), the amplitudes can be represented in the form

$$w^{(\mathbf{k})}=\sqrt{\frac{E_0\varepsilon_{\mathbf{n}}}{L}}\cdot\frac{1}{\left|\Xi^{(\mathbf{k})}\right|}\sum_{\mathbf{p}\in L^{(\mathbf{k})}}\xi_{\mathbf{p}}^{(\mathbf{k})*}w_{\mathbf{p}} \tag{3.8}$$

Equation (3.8) defines a linear combination of modal coefficients $w_{\mathbf{p}}$.

Taking $\xi_{\mathbf{p}}^{(\mathbf{k})}=\delta_{\mathbf{pk}}$ gives us a possibility to obtain in each branch the complex amplitude $w^{(\mathbf{p})}$ proportional to one modal coefficient $w_{\mathbf{p}}$, and to measure the modal power $\sim\left|w_{\mathbf{p}}\right|^2$ using a photoreceiver.

Using the more complicated weight coefficients

$$\xi_{\mathbf{p}}^{(\mathbf{k}_c)}=\delta_{\mathbf{p}_1\mathbf{k}}+\delta_{\mathbf{p}_2\mathbf{k}}$$

$$\xi_{\mathbf{p}}^{(\mathbf{k}_s)}=\delta_{\mathbf{p}_1\mathbf{k}}+\exp(i\Delta\varphi)\,\delta_{\mathbf{p}_2\mathbf{k}},\quad \Delta\varphi=const$$

leads to the formation of the complex coefficients

$$w^{(\mathbf{k}_c)}=w_{\mathbf{p}_1}+w_{\mathbf{p}_2} \tag{3.9}$$

$$w^{(\mathbf{k}_s)}=w_{\mathbf{p}_1}+\exp(i\Delta\varphi)\,w_{\mathbf{p}_2} \tag{3.10}$$

in the branches numbered $\mathbf{k}_c$ and $\mathbf{k}_s$, respectively.

A quadratic point-aperture photo-receiver measures the interference of the coefficients $w_{\mathbf{p}_1}$ and $w_{\mathbf{p}_2}$ in the points $\lambda f_0\boldsymbol{\nu}^{(\mathbf{k}_c)}$, $\lambda f_0\boldsymbol{\nu}^{(\mathbf{k}_s)}$ that enables us to reconstruct the phases $\varphi_{\mathbf{p}} \equiv \arg w_{\mathbf{p}}$ of the coefficients $w_{\mathbf{p}}$. Thereby a modal composition is measured, which includes both the modal power and the intermode phase shifts. The phase $\arg w_{\mathbf{p}}$ reconstruction is carried out by a set of equations similar to that for processing of the quadrature interferograms. To obtain the required equations, let us write Eqs.(3.9) and (3.10) as follows

$$\left|w^{(\mathbf{k}_c)}\right|^2 \sim \left|w_{\mathbf{p}_1}\right|^2 + \left|w_{\mathbf{p}_2}\right|^2 + 2\left|w_{\mathbf{p}_1}\right|\left|w_{\mathbf{p}_2}\right|\cos\left(\varphi_{\mathbf{p}_1} - \varphi_{\mathbf{p}_2}\right)$$

$$\left|w^{(\mathbf{k}_c)}\right|^2 \sim \left|w_{\mathbf{p}_1}\right|^2 + \left|w_{\mathbf{p}_2}\right|^2 + 2\left|w_{\mathbf{p}_1}\right|\left|w_{\mathbf{p}_2}\right| \times \left[\cos\left(\varphi_{\mathbf{p}_1} - \varphi_{\mathbf{p}_2}\right)\cos\Delta\varphi - \sin\left(\varphi_{\mathbf{p}_1} - \varphi_{\mathbf{p}_2}\right)\sin\Delta\varphi\right] \tag{3.11}$$

Assume that the power $\left|w_{\mathbf{p}_1}\right|^2$ is measured in the branch number $\mathbf{k}_1$ and the power $\left|w_{\mathbf{p}_2}\right|^2$ in the branch number $\mathbf{k}_2$, so that

$$\left|w^{(\mathbf{k}_1)}\right|^2 \sim \left|w_{\mathbf{p}_1}\right|^2, \quad \left|w^{(\mathbf{k}_2)}\right|^2 \sim \left|w_{\mathbf{p}_2}\right|^2 \tag{3.12}$$

In such a case the power of modes $\mathbf{p}_1$ and $\mathbf{p}_2$ can be found as a result of measurements of Eq.(3.12). Then, to reconstruct the phase difference, $\varphi_{\mathbf{p}_1} - \varphi_{\mathbf{p}_2}$, it proves to be sufficient to solve Eqs.(3.11) and (3.12) with respect to the cosine and sine of that difference:

$$\cos\left(\varphi_{\mathbf{p}_1} - \varphi_{\mathbf{p}_2}\right) = \frac{\left|w^{(\mathbf{k}_c)}\right|^2 - \left|w^{(\mathbf{k}_1)}\right|^2 - \left|w^{(\mathbf{k}_2)}\right|^2}{2\left[\left|w^{(\mathbf{k}_1)}\right|^2\left|w^{(\mathbf{k}_2)}\right|^2\right]^{1/2}}$$

$$\sin\left(\varphi_{\mathbf{p}_1} - \varphi_{\mathbf{p}_2}\right) = \frac{\cos\left(\varphi_{\mathbf{p}_1} - \varphi_{\mathbf{p}_2}\right)}{\operatorname{tg}\Delta\varphi} - \frac{\left|w^{(\mathbf{k}_s)}\right|^2 - \left|w^{(\mathbf{k}_1)}\right|^2 - \left|w^{(\mathbf{k}_2)}\right|^2}{2\sin\Delta\varphi\left[\left|w^{(\mathbf{k}_1)}\right|^2\left|w^{(\mathbf{k}_2)}\right|^2\right]^{1/2}}$$

Then, taking into account the sign of $\cos\left(\varphi_{\mathbf{p}_1} - \varphi_{\mathbf{p}_2}\right)$, the difference $\varphi_{\mathbf{p}_1} - \varphi_{\mathbf{p}_2}$ can be derived from the equation

$$\operatorname{tg}\left(\varphi_{\mathbf{p}_1} - \varphi_{\mathbf{p}_2}\right) = \frac{1}{\operatorname{tg}\Delta\varphi} - \frac{1}{\sin\Delta\varphi}\,\frac{\left|w^{(\mathbf{k}_s)}\right|^2 - \left|w^{(\mathbf{k}_1)}\right|^2 - \left|w^{(\mathbf{k}_2)}\right|^2}{\left|w^{(\mathbf{k}_c)}\right|^2 - \left|w^{(\mathbf{k}_1)}\right|^2 - \left|w^{(\mathbf{k}_2)}\right|^2}.$$

We can find the proper values of the phase $\varphi_{\mathbf{p}}$ either in a recurrent manner from the $\varphi_{\mathbf{p}_1} - \varphi_{\mathbf{p}_2}$ difference or with respect to one basic branch $\mathbf{p}_0$. Note that for $\Delta\varphi = \frac{\pi}{2}$ the method is equivalent to that proposed by [Kotlyar, 1989].

## 3.4. MODAL SPECTRAL TRANSFORMS

It is in the process of design of fiber-optical communication systems that a topical problem arises of correcting for the mode differential attenuation and the differential modal delay since they cause the blurring of the impulse form. Speaking mathematically, we must perform here the alteration of the amplitude and phase of the coefficients of the light beam expansion in terms of modes, that is, the mode filtration. The problem of mode filtration also arises when we need to form a time-dependent light beam for using in a fiber-optical communication line with the modal multiplexing of channels. In the last case we must perform the time-dependent alteration of separate modes and be able to sum up modes keeping pace with data transmission.

For the sinusoidal modes in Eq.(1.112), there is a conventional system of optical spatial filtration that consists of two Fourier-stages (those for direct and inverse transforms) and a complex spatial filter in the Fourier plane known also as a plane of spatial frequencies, which is placed between the stages. Such a system allows us to perform the transform of spatial spectrum. If we could have filtration systems suitable for more general kinds of modes, then the possibility to control the light beam modal composition would be opened. In particular, we could perform time modulation of mode coefficients, which proves to be necessary for fiber-optical communication lines that use the modal channel multiplexing.

However as far as arbitrary (non-sinusoidal) mode functions are involved, it is not a simple matter to design an optical system capable of doing both direct and inverse modal compositions. However, we can here employ a stage for the mode analysis with multiple carrier frequencies to perform a direct transform and a similar stage for modal formation for doing an inverse transform. In this case, we shall make use of a sub-group of singlemodal multibeam modans intended for forming a singlemodal beam in each branch.

The task of the mode filtration can be solved if we develop a two-stage optical system with the modans $M_1$ and $M_2$ that includes the stages for doing the direct and inverse mode transform (Fig. 3.6). A plane, which is common for the output of the direct and the input of the inverse mode transforms, is similar to the Fourier plane of a coherent-optical correlator and can be referred to as a modal plane or a generalized-spectral plane. Each point in a modal plane that is placed in a discrete grid, corresponds to one mode so that the change of a modal composition can be carried out by using a conventional complex spatial filter $\Delta\Gamma$.

Modan $M_1$ must restore several images in several angular separated beams. The purpose of the modan $M_2$ (Fig. 3.6) here is to superimpose several beams tilted at different angles. Given an appropriate angular selectivity of the transfer (which is typical of 3-D holographic elements, see [Denisyuk, 1986; Wang, 1990], the modan can work efficiently. Actually, the modan must be a 3-D hologram that restores tilted beams corresponding to discrete spatial frequencies. However, a thin modan can work as $M_2$ with less energy efficiency that 3-D. The summation is performed by this modan in several diffraction orders in parallel.

If we place into a modal plane a transparency with the function of the complex transfer $\Delta\Gamma(\mathbf{v})$, then the modal coefficients $w_{\mathbf{p}}$ of the input field $w(\mathbf{x})$ will be transformed into the modal coefficients $W_{\mathbf{p}}$ of the output field $W(\mathbf{x})$ in accordance with the equations

$$W_{\mathbf{p}} = \gamma_{\mathbf{p}} \cdot w_{\mathbf{p}}, \quad \mathbf{p} \in K, \; \gamma_{\mathbf{p}} = \Delta\Gamma(\mathbf{v}^{(\mathbf{p})}).$$

The $W_{\mathbf{p}}$ coefficients are then used in the stage for inverse mode transform to form the output filtered field.

Let us discuss optical stages for the direct and inverse mode transforms in more detail. The direct transform can be treated as the conversion of the field in the beam cross-section, Eq.(3.3), to the mode coefficients of Eq.(3.5) which are ordered, spatially separated, and imaged as pixels of the complex amplitude in the corresponding points of an output plane (Fig. 3.4).This is the case when a set of the measured modal coefficients $\{w_{\mathbf{p}}, \mathbf{p} \in K\}$ should be complete in terms of the definition of the field $w(\mathbf{x})$. For example, the completeness can be regarded in the sense of the approximation accuracy

$$w(\mathbf{x}) \cong \sum_{\mathbf{p}\in K} w_{\mathbf{p}} \psi_{\mathbf{p}}(\mathbf{x})$$

in accordance with the relative mean-square error criterion:

$$\delta_0^2 = 1 - \frac{1}{E} \sum_{\mathbf{p}\in K} \left|w_{\mathbf{p}}\right|^2, \tag{3.13}$$

where $E = \int_G \left|w(\mathbf{x})\right|^2 d^2\mathbf{x}$ is the power of the input beam.

Note that the $\delta_0^2$ magnitude substantially depends on the power distribution between modes $\left|w_{\mathbf{p}}\right|^2 / E$ that occurs in the beam under study $w(\mathbf{x})$, see Eq.(3.3). Being some kind of modal analysis, the direct modal transform is very specific case of that analysis. The main feature that must be emphasized is the use of all the modal coefficients (3.5) or the multimodal field (3.3). For the sake of comparison, note that earlier in section 3.3, we were studying several (may be one) modal coefficient, Eq.(3.5), of the field under study.

Let us consider how to perform the optical modal transform of the field $w(\mathbf{x})$ by means of a Fourier stage at whose input plane a singlemodal multibeam modan (Fig. 3.4) is placed. If the modan $M_1$ is illuminated by plane wave, it must reconstruct the beam $t^*(\mathbf{x})$, where $t(\mathbf{x})$ is specified by Eq.(2.47) upon the plane illuminating beam and with

$$\mu^{(\mathbf{p})} \equiv \frac{1}{\sqrt{L}}, \tag{3.14}$$

The spatial frequencies are to be placed on a rectangular grid $\boldsymbol{\nu}^{(\mathbf{p})} = \left(\nu_x^{(p)}, \nu_y^{(l)}\right)$, $p, l = 0, \pm 1, \pm 2, \ldots$ characterized by constant steps along the axes, i.e.,

$$\nu_x^{(p)} = p\nu_x^1, \quad \nu_y^{(l)} = l\nu_y^1 \tag{3.15}$$

A screen with pinholes that are centered at $\boldsymbol{\nu}^{(\mathbf{p})}$, must be put at the output plane. Special diffusers can be used instead of the pinhole (see [Lee, 1974]).Generating functions (see section 2.4) with the carrier frequencies (3.15) turn out to be especially useful for the computation of singlemodal modans.

Now treating an optical system of Fig. 3.4 as particular case of the modal analyzer, we find it from Eqs.(3.8) and (3.14) that the modan ensures the calculating of the coefficients

$$w^{(\mathbf{p})} = \sqrt{\frac{E_0 \varepsilon_{\mathbf{n}}}{L}} w_{\mathbf{p}} \tag{3.16}$$

in the discrete points of the Fourier plane where $w_{\mathbf{p}}$ is specified by Eq.(3.5).

Thus, the stage for the direct mode transform can be realized on the basis of a thin holographic optical element coupled with a Fourier stage.

Performing of an inverse mode transform, i.e., excitation the multimodal field $W(\mathbf{x})$ from the given set of its modal coefficients $\{W_{\mathbf{p}}, \mathbf{p} \in K\}$, implies excitation such a complete set of modes $\{\psi_{\mathbf{p}}, \mathbf{p} \in K\}$ that the result of mode superposition

$$\sum_{\mathbf{p} \in K} W_{\mathbf{p}} \psi_{\mathbf{p}}(\mathbf{x}) \cong W(\mathbf{x})$$

should approximate the field $W(\mathbf{x})$, with respect to the mean square criterion

$$\delta_\gamma^2 = 1 - \frac{1}{E_\gamma} \sum_{\mathbf{p} \in K} \left| W_\mathbf{p} \right|^2 \tag{3.17}$$

Here $E_\gamma = \int_G |W(\mathbf{x})|^2 d^2\mathbf{x}$ is the power of the output beam.

We shall perform an inverse modal transform given by the equation

$$W(\mathbf{x}') \cong \sum_{\mathbf{p} \in K} W_\mathbf{p} \psi_\mathbf{p}(\mathbf{x}')$$

using an inverse Fourier stage (Fig. 3.5) that has in its output plane a singlemodal multibeam modan. Modan here is thin holographic optical element that reconstructs the beam $t(\mathbf{x})$. The inverse Fourier stage has in its input plane a screen with a set of pinholes centered at $\boldsymbol{\nu}^{(\mathbf{p})}$. If we create in the $\boldsymbol{\nu}^{(\mathbf{p})}$points the complex amplitude proportional to the mode coefficients $W_\mathbf{p}$, this yields us a matrix of complex sources in the form

$$\sum_{\mathbf{p} \in K} W_\mathbf{p} \delta\left(\boldsymbol{\nu} - \boldsymbol{\nu}^{(\mathbf{p})}\right),$$

where δ(.) is a 2-D Dirac $\delta$-function.

Directly in front of the modan plane the following beam is formed

$$\sum_{\mathbf{p} \in K} W_\mathbf{p} \exp\left(i 2\pi \boldsymbol{\nu}^{(\mathbf{p})} \mathbf{x}'\right)$$

that is multiplied by the complex function of the modan transfer determined (provided that $w_0(\mathbf{x}) = const$) by

$$\Gamma_\mathbf{n}(\mathbf{x}') = \frac{1}{w_0} \sqrt{\frac{E_0 \varepsilon_\mathbf{n}}{L}} \sum_{\mathbf{p} \in K} \psi_{\mathbf{p}'}(\mathbf{x}') \exp\left(i 2\pi \boldsymbol{\nu}^{(\mathbf{p}')} \mathbf{x}'\right) \exp\left(i 2\pi \boldsymbol{\nu}_\mathbf{n} \mathbf{x}'\right). \tag{3.18}$$

Thus, the field directly behind the modan plane is equal

$$\begin{aligned} &\sqrt{\frac{E\varepsilon_\mathbf{n}}{L}} \sum_{\mathbf{p} \in K} W_\mathbf{p} \left\{ \psi_\mathbf{p}(\mathbf{x}') + \sum_{\mathbf{p}' \neq \mathbf{p}} \psi_{\mathbf{p}'}(\mathbf{x}') \exp\left[i 2\pi \left(\boldsymbol{\nu}^{(\mathbf{p}')} - \boldsymbol{\nu}^{(\mathbf{p})}\right) \mathbf{x}'\right] \right\} \times \\ &\times \exp\left[i 2\pi \left(\boldsymbol{\nu}_\mathbf{n} + \boldsymbol{\nu}^{(\mathbf{p})}\right) \mathbf{x}'\right] = \\ &= \sqrt{\frac{E\varepsilon_\mathbf{n}}{L}} \left\{ \sum_{\mathbf{p} \in K} W_\mathbf{p} \psi_\mathbf{p}(\mathbf{x}') + \sum_{\mathbf{p} \in K} W_\mathbf{p} \sum_{\mathbf{p}' \neq \mathbf{p}} \psi_{\mathbf{p}'}(\mathbf{x}') \exp\left[i 2\pi \left(\boldsymbol{\nu}^{(\mathbf{p}')} - \boldsymbol{\nu}^{(\mathbf{p})}\right) \mathbf{x}'\right] \right\} \times \\ &\times \exp\left[i 2\pi \left(\boldsymbol{\nu}_\mathbf{n} + \boldsymbol{\nu}^{(\mathbf{p})}\right) \mathbf{x}'\right] \end{aligned} \tag{3.19}$$

Since $\boldsymbol{\nu}^{(\mathbf{p}')} - \boldsymbol{\nu}^{(\mathbf{p})} \neq 0$ for $\mathbf{p}' \neq \mathbf{p}$, the terms of the second sum have the frequencies that differ from the frequency of the first sum by at least $\nu_x^{(1)}$ on the $x$-axis and/or by $\nu_y^{(1)}$ on the $y$-axis, i.e., they can be spatially separated if

$$\nu_x^{(1)} \geq \frac{1}{2a}, \quad \nu_y^{(1)} \geq \frac{1}{2b},$$

where $2a \times 2b$ is the size of the region $G$ of a modan.

With the orders separated, the power of the field of Eq.(3.19) can be calculated by the formula

$$\frac{E\varepsilon_{\mathbf{n}}}{L} \sum_{\mathbf{p} \in K} \left| W_{\mathbf{p}} \right|^2$$

and appears to be only a part of the input power. Thus, the input power is divided into $L$ portions, with only the $1/L$ portion utilized. Therefore, there are the energy losses, which are due to the multibeam modan that is used and provides summation of the mode functions in each branch with the corresponding separation of light power into portions.

# Chapter 4

# MODAN'S PERFORMANCES

## 4.1. PERTURBATIONS OF MODAL FUNCTIONS

A modan dedicated to performing the required modal transform is characterized by the continuous function of the complex transfer $T$ in the form of Eqs.(2.32) and (2.2). When the modan represents an optical element, which is fabricated as holographic optical element by the computer-aided techniques, there arise a variety of perturbations of the complex transfer function. The perturbations are due to (a) the coding of the complex function into a phase-only form (see section 2.5), (b) the sampling taking place in computer-aided design, (c) the form of the mask writer raster spot, which is strictly determined and (d) a variety of errors in the technology for the formation of the diffraction microrelief.

When modans are realized by means of the computer generation of holograms, both the perturbations of quantization and the sampling of transfer function turn out to be most specific.

A general property of all perturbations is substitution of the coded transfer function $\Gamma(\mathbf{x})$ of a modan by a perturbed transfer function

$$\hat{\Gamma}(\mathbf{x}) = \hat{D}\Gamma(\mathbf{x}),$$

where $\hat{D}$ is an operator describing the foregoing perturbations and supposed to be linear here. The perturbed transfer function

$$\hat{D}\Gamma_{\mathbf{n}}(\mathbf{x}) = g_{\mathbf{n}}\hat{D}\left[T(\mathbf{x})\exp(i2\pi\mathbf{v}_{\mathbf{n}}\mathbf{x})\right].$$

operates in the working diffraction order of a modan in the presence of perturbations due to Eq.(2.42)

Reconstructed light beam in the $\mathbf{n}$ -th diffraction order is described by the equation

$$\hat{\ }_{\mathbf{n}}(\mathbf{x}) = g_{\mathbf{n}}c\sqrt{E_0}\,w_0(\mathbf{x})\hat{D}\left[\frac{t(\mathbf{x})}{w_0(\mathbf{x})}\exp(i2\pi\mathbf{n}_{\mathbf{n}}\mathbf{x})\right], \tag{4.1}$$

for the illuminating beam $w_0(\mathbf{x})$, and with Eq.(2.32) taken into account, where the constant $c$ is specified by Eq.(2.33). Owing to Eqs.(2.2) and (2.6), the function $\hat{W}_{\mathbf{n}}$ of Eq.(4.1) may be represented in the form

$$\hat{W}_{\mathbf{n}} = \sum_{\mathbf{k}} \hat{W}_{\mathbf{n}}^{(\mathbf{k})}, \tag{4.2}$$

which is a superposition of partial beams $\hat{W}_{\mathbf{n}}^{(\mathbf{k})}$ propagating in their branches and specified by the expression

$$\hat{W}_{\mathbf{n}}^{(\mathbf{k})}(\mathbf{x}) = \sqrt{E_0} g_{\mathbf{n}} C \mu^{(\mathbf{k})} \frac{\eta^{(\mathbf{k})}(\mathbf{x})}{\left|\Xi^{(\mathbf{k})}\right|} \exp\left(i 2\pi \mathbf{v}_{\mathbf{n}}^{(\mathbf{k})} \mathbf{x}\right), \tag{4.3}$$

Here functions

$$\eta^{(\mathbf{k})}(\mathbf{x}) = \sum_{\mathbf{p} \in L^{(\mathbf{k})}} \xi_{\mathbf{p}}^{(\mathbf{k})} \varphi_{\mathbf{p}}^{(\mathbf{k})}(\mathbf{x}) \tag{4.4}$$

are the "perturbed" mode beams expressed in terms of "perturbed" base functions

$$\varphi_{\mathbf{p}}^{(\mathbf{k})}(\mathbf{x}) = w_0(\mathbf{x}) \exp\left(-i 2\pi \boldsymbol{\nu}_{\mathbf{n}}^{(\mathbf{k})} \mathbf{x}\right) \cdot \hat{D}\left\{\frac{\psi_{\mathbf{p}}(\mathbf{x})}{w_0(\mathbf{x})} \exp\left(i 2\pi \boldsymbol{\nu}_{\mathbf{n}}^{(\mathbf{k})} \mathbf{x}\right)\right\} \tag{4.5}$$

Equations (4.3) and (4.5) may be interpreted as follows: even though there are perturbations, we can treat one branch of modan as the certain direction in space that is characterized by the spatial frequency $\boldsymbol{\nu}_{\mathbf{n}}^{(\mathbf{k})}$, Eq.(2.48), and by the unit vector $\mathbf{N}_{\mathbf{n}}^{(\mathbf{k})}$.

According to Eqs.(4.3) and (2.47), the presence of perturbations results in the replacement of the mode beams $\xi^{(\mathbf{k})}(\mathbf{x})$, in Eq.(2.47) by the corresponding mode beams $\eta^{(\mathbf{k})}(\mathbf{x})$ to obtain Eq.(4.3).

According to Eqs.(2.6), (4.4), the perturbed functions $\varphi_{\mathbf{p}}^{(\mathbf{k})}(\mathbf{x})$ occur in the **k**-th branch instead of the modes $\psi_{\mathbf{p}}(\mathbf{x})$, as a result of perturbation of a modans' complex transfer function. Let us introduce perturbations as the functions (see Fig. 4.1)

$$h_{\mathbf{p}}^{(\mathbf{k})}(\mathbf{x}) = \varphi_{\mathbf{p}}^{(\mathbf{k})}(\mathbf{x}) - \psi_{\mathbf{p}}(\mathbf{x}), \quad \mathbf{x} \in G \tag{4.6}$$

reported in the work [Golub, 1990b].

Each perturbed function $\varphi_{\mathbf{p}}^{(\mathbf{k})}(\mathbf{x})$, is similar to the modes and can be interpreted as a result of interaction between the perturbation $h_{\mathbf{p}}(\mathbf{x})$ and the orthonormal mode function $\psi_{\mathbf{p}}(\mathbf{x})$.

The perturbations introduced in Eqs.(4.5) and (4.6) take into account the forms of the modes $\psi_{\mathbf{p}}(\mathbf{x})$ and the presence of spatial carrier frequencies $\boldsymbol{\nu}_{\mathbf{n}}^{(\mathbf{k})}$.

It is convenient to describe perturbations by the means of matrix elements

$$H_{\mathbf{pp}'}^{(\mathbf{k})} = \int_G h_{\mathbf{p}}^{(\mathbf{k})*}(\mathbf{x}) \psi_{\mathbf{p}'}(\mathbf{x}) \rho_0(\mathbf{x}) d^2\mathbf{x} = \left(h_{\mathbf{p}}^{(\mathbf{k})}, \psi_{\mathbf{p}'}\right)^* \tag{4.7}$$

and quadratic matrix elements

$$Q_{\mathbf{pp}'}^{(\mathbf{k},\mathbf{k}')} = \int_G h_{\mathbf{p}}^{(\mathbf{k})*}(\mathbf{x}) h_{\mathbf{p}'}^{(\mathbf{k}')}(\mathbf{x}) \rho_0(\mathbf{x}) d^2\mathbf{x} = \left(h_{\mathbf{p}}^{(\mathbf{k})}, h_{\mathbf{p}'}^{(\mathbf{k}')}\right)^* \tag{4.8}$$

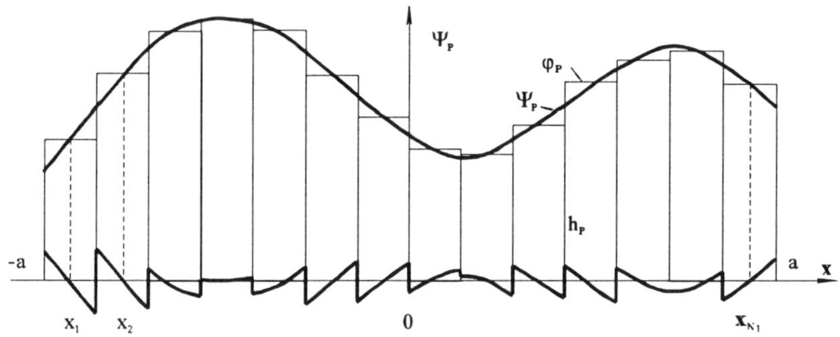


**Figure 4.1.** A mode function perturbation upon sampling. (Modified from Golub M.A., Sisakyan I.N., Soifer V.A., Computer generated optical elements for fiber's mode selection and launching, *Proceedings SPIE, OE/Fibers'90*, 1365, 156, 1990).

"In-branch" matrices of perturbations defined as

$$H^{(\mathbf{k})}=\left[H^{(\mathbf{k})}_{\mathbf{pp}'}\right],\ Q^{(\mathbf{k})}=\left[Q^{(\mathbf{k},\mathbf{k})}_{\mathbf{pp}'}\right],\ \mathbf{p},\mathbf{p}'\in L^{(\mathbf{k})},\mathbf{k}\in K. \tag{4.9}$$

will be very useful for compact notations

Note that the perturbation parameters may differ from branch to branch of the modan due to the carrier $\boldsymbol{\nu}^{(\mathbf{k})}_{\mathbf{n}}$ varying with respect to $\mathbf{k}$. Perturbation effects described by Eq.(4.5) lead both to the redistribution of the power and phase between branch as well as to the redistribution of the power and phase between the modes in each branch. The perturbed functions $\varphi^{(\mathbf{k})}_{\mathbf{p}}$ are related to the modes by the expression

$$\varphi^{(\mathbf{k})}_{\mathbf{p}}(\mathbf{x})=\psi_{\mathbf{p}}(\mathbf{x})+\sum_{\mathbf{p}'}H^{(\mathbf{k})*}_{\mathbf{pp}'}\psi_{\mathbf{p}'}(\mathbf{x}), \tag{4.10}$$

that is simply an expansion in terms of the orthonormal functions $\{\psi_{\mathbf{p}}\}$.

In Eq.(4.10) a summation is carried out over all values of **p'**. It is seen that the useful mode $\psi_{\mathbf{p}}$ in the **p**-th branch is present with an extra complex-valued factor $1+H^{(\mathbf{k})^*}_{\mathbf{pp}}$ that generates an approximation in the form

$$\psi_{\mathbf{p}}(\mathbf{x})=\frac{\varphi^{(\mathbf{k})}_{\mathbf{p}}(\mathbf{x})}{1+H^{(\mathbf{k})*}_{\mathbf{pp}}}. \tag{4.11}$$

The power of the required mode $\psi_{\mathbf{p}}$ is equal to

$$E_{\mathbf{p}}=\left|1+H^{(\mathbf{k})}_{\mathbf{pp}}\right|^2, \tag{4.12}$$

The power of the parasitic mode $\psi_{\mathbf{p}'}$ $(\mathbf{p}' \neq \mathbf{p})$, upon excitation of the **p**-th mode, equals $\left|H_{\mathbf{pp}'}^{(\mathbf{k})}\right|^2$. A total power of parasitic modes specified with the help of the Parseval equality is given by the equation

$$\Delta E_{\mathbf{p}} = \sum_{\mathbf{p}' \neq \mathbf{p}} \left|H_{\mathbf{pp}'}^{(\mathbf{k})}\right|^2 = \left\|h_{\mathbf{p}}^{(\mathbf{k})}\right\|^2 - \left|H_{\mathbf{pp}'}^{(\mathbf{k})}\right|^2 = Q_{\mathbf{pp}}^{(\mathbf{k},\mathbf{k})} - \left|H_{\mathbf{pp}'}^{(\mathbf{k})}\right|^2 \tag{4.13}$$

A total power of the beam from the sampled modan equals to a sum

$$E_{\mathbf{p}} + \Delta E_{\mathbf{p}} = 1 + Q_{\mathbf{pp}}^{(\mathbf{k},\mathbf{k})} + 2\,\mathrm{Re}\, H_{\mathbf{pp}}^{(\mathbf{k})}$$

that is near to the value $1 \equiv \left\|\psi_{\mathbf{p}}\right\|^2$.

It is important and also practical to consider here a special case of single-modal modans. For single-modal modans only one perturbation acts in each branch number **p**

$$h_{\mathbf{p}} = h_{\mathbf{p}}^{(\mathbf{p})} \tag{4.14}$$

The perturbation is characterized by one string of matrix elements $H_{\mathbf{pp}'}^{(\mathbf{p})}$, see Eq.(4.7). The matrixes of the in-branch perturbations, Eq.(4.9), degenerate into the single numbers

$$H^{(\mathbf{p})} \equiv H_{\mathbf{pp}}^{(\mathbf{p})} = \left(h_{\mathbf{p}}, \psi_{\mathbf{p}}\right)^* \tag{4.15}$$

$$Q^{(\mathbf{p})} \equiv Q_{\mathbf{pp}}^{(\mathbf{p},\mathbf{p})} = \left\|h_{\mathbf{p}}\right\|^2 = \sum_{\mathbf{p}'} \left|H_{\mathbf{pp}'}^{(\mathbf{p})}\right|^2. \tag{4.16}$$

Approximation of Eq.(4.11) and expansion of Eq.(4.10) make it possible to write down

$$\frac{\varphi_{\mathbf{p}}(\mathbf{x})}{1 + H^{(\mathbf{p})*}} - \psi_{\mathbf{p}} = \frac{h_{\mathbf{p}} - H^{(\mathbf{p})*}\psi_{\mathbf{p}}}{1 + H^{(\mathbf{p})*}}. \tag{4.17}$$

Using the designations from Eqs.(4.15) and (4.16), we can write down the expression for the error of the approximation of Eq.(4.11) as follows

$$\left\|\frac{\varphi_{\mathbf{p}}(\mathbf{x})}{1 + H^{(\mathbf{p})*}} - \psi_{\mathbf{p}}\right\|^2 = \frac{Q^{(\mathbf{p})} - \left|H^{(\mathbf{p})}\right|^2}{\left|1 + H^{(\mathbf{p})}\right|^2}. \tag{4.18}$$

It seems expedient to introduce the inter-branch matrices of perturbation for singlemodal modans:

$$H_L = \left[H_{\mathbf{pp}'}^{(\mathbf{p})};\ \mathbf{p},\mathbf{p}' \in K\right],\quad H_{\mathbf{pp}'}^{(\mathbf{p})} = \left(h_{\mathbf{p}}, \psi_{\mathbf{p}'}\right)^* \tag{4.19}$$

$$Q_L = \left[Q_{\mathbf{pp}'}^{(\mathbf{p},\mathbf{p}')};\ \mathbf{p},\mathbf{p}' \in K\right],\ \ Q_{\mathbf{pp}'}^{(\mathbf{p},\mathbf{p}')} = \left(h_{\mathbf{p}}, h_{\mathbf{p}'}\right)^* \tag{4.20}$$

In the special case of single-modal modans under considerations we are introducing the notations

$$H_r^{(\mathbf{p})} = -\left(-H^{(\mathbf{p})^*}\right)^{r+1};\ \ R_r^{(\mathbf{p})} = H_r^{(\mathbf{p})} + H_r^{(\mathbf{p})^*} + \left|H_r^{(\mathbf{p})}\right|^2 \tag{4.21}$$

useful for the foregoing considerations.

## 4.2. MODAL COMPOSITION IN PRESENCE OF PERTURBATIONS

### A. Decrease of the modan's selectivity caused by perturbations.

Decrease of the modan's selectivity caused by perturbations shows itself through the two factors:

1) the change of the power distribution and of the phase shift between the required modes numbered $\mathbf{p} \in L^{(\mathbf{k})}$.

2) the excitation of "parasitic" modes numbered $\mathbf{p} \notin L^{(\mathbf{k})}$ that occurs at the expense of the portion of the power of the required modes.

The perturbed function in Eq.(4.4) for a beam described by Eq.(4.3) can be represented in the form

$$\eta^{(\mathbf{k})}(\mathbf{x}) = \eta_0^{(\mathbf{k})}(\mathbf{x}) + \Delta\eta^{(\mathbf{k})}(\mathbf{x}). \tag{4.22}$$

Note that by virtue of Eq.(4.11), the beam

$$\eta_0^{(\mathbf{k})}(\mathbf{x}) = \sum_{\mathbf{p}\in L^{(\mathbf{k})}} \eta_{\mathbf{p}}^{(\mathbf{k})} \psi_{\mathbf{p}}(\mathbf{x}) \tag{4.23}$$

corresponds to the required modes with the coefficients

$$\eta_{\mathbf{p}}^{(\mathbf{k})} = \xi_{\mathbf{p}} + \sum_{\mathbf{p}'\in L^{(\mathbf{k})}} H_{\mathbf{pp}'}^{(\mathbf{k})*} \xi_{\mathbf{p}'}^{(\mathbf{k})},\ \ \mathbf{p} \in L^{(\mathbf{k})} \tag{4.24}$$

that form a vector of the "perturbed" mode composition:

$$\mathrm{H}^{(\mathbf{k})} = \left[E^{(\mathbf{k})} + H^{(\mathbf{k})}\right]\Xi^{(\mathbf{k})}, \tag{4.25}$$

where $E^{(\mathbf{k})}$ is the unit matrix measured $\left|L^{(\mathbf{k})}\right| \times \left|L^{(\mathbf{k})}\right|$.

Parasitic modes produce a beam

$$\Delta\eta^{(\mathbf{k})}(\mathbf{x}) = \sum_{\mathbf{p}\in L^{(\mathbf{k})}} \eta_{\mathbf{p}}^{(\mathbf{k})} \psi_{\mathbf{p}}(\mathbf{x}) \tag{4.26}$$

where

$$\eta_{\mathbf{p}}^{(\mathbf{k})} = \sum_{\mathbf{p}'\in L^{(\mathbf{k})}} H_{\mathbf{pp}'}^{(\mathbf{k})*} \xi_{\mathbf{p}'}^{(\mathbf{k})},\ \ \mathbf{p} \notin L^{(\mathbf{k})}. \tag{4.27}$$

Owing to Parseval equality we find that

$$\left\|\eta^{(\mathbf{k})}\right\|^2 = \sum_{\mathbf{p}} \left|\eta_{\mathbf{p}}^{(\mathbf{k})}\right|^2 = \left\|\eta_0^{(\mathbf{k})}\right\|^2 + \left\|\Delta\eta^{(\mathbf{k})}\right\|^2, \tag{4.28}$$

where

$$\left\|\eta_0^{(\mathbf{k})}\right\|^2 = \sum_{\mathbf{p}\in L^{(\mathbf{k})}} \left|\eta_{\mathbf{p}}^{(\mathbf{k})}\right|^2 = \left|\mathrm{H}^{(\mathbf{k})}\right|^2, \tag{4.29}$$

$$\left\|\Delta\eta^{(\mathbf{k})}\right\|^2 = \sum_{\mathbf{p}\notin L^{(\mathbf{k})}} \left|\eta_{\mathbf{p}}^{(\mathbf{k})}\right|^2 \tag{4.30}$$

**B. Correction for perturbations**

In order to achieve the power and phase distributions between the modes nearest to the required ones (despite the perturbation effect) we suggest to substitute original coefficients $\Xi^{(\mathbf{k})}$ Eq.(2.6) by the predistorted coefficients designated as

$$\tilde{\Xi}^{(\mathbf{k})} = \left(\tilde{\xi}_{\mathbf{p}}^{(\mathbf{k})}, \mathbf{p} \in L^{(\mathbf{k})}\right), \quad \mathbf{k} \in K$$

Let the condition of small perturbations

$$\left\|H^{(\mathbf{k})}\right\| < 1$$

be satisfied. Owing to Eq.(4.25), we can represent the vector $\tilde{\Xi}^{(\mathbf{k})}$ of predistorted coefficients in the form of the Neimann series see [Ahiezer, 1966] whose first $r$ terms give us

$$\tilde{\Xi}^{(\mathbf{k})} = \sum_{r'=0}^{r} \left(-H^{(\mathbf{k})*}\right)^{r'} \Xi^{(\mathbf{k})}. \tag{4.31}$$

Let us represent Eq.(4.31) in the recurrence form

$$\tilde{\Xi}_0^{(\mathbf{k})} = \Xi^{(\mathbf{k})}$$

$$\tilde{\Xi}_{r'}^{(\mathbf{k})} = \left(-H^{(\mathbf{k})*}\right)\tilde{\Xi}_{r'-1}^{(\mathbf{k})} + \Xi^{(\mathbf{k})}, \quad r' = \overline{1,r} \tag{4.32}$$

$$\tilde{\Xi}^{(\mathbf{k})} = \tilde{\Xi}_r^{(\mathbf{k})}$$

and refer to these equations as the correction operation of the $r$-th order. After performing this, we obtain a vector of the perturbed mode coefficients that, allowing for Eq.(4.3), can be written as follows

$$\mathrm{H}^{(\mathbf{k})} = \left[E^{(\mathbf{k})} + H^{(\mathbf{k})*}\right]\tilde{\Xi}^{(\mathbf{k})} = \left[E^{(\mathbf{k})} + H_r^{(\mathbf{k})}\right]\Xi^{(\mathbf{k})}, \tag{4.33}$$

where

$$H_r^{(\mathbf{k})} = \left[-H^{(\mathbf{k})*}\right]^{r+1}. \qquad (4.34)$$

Matrix $H_r^{(\mathbf{k})}$ becomes simply a value (4.21) in the case of single-modal modans.

It is essential that introducing the correction operation of the $r$-th arbitrary order permits us to find unified relationships that cover both the limiting case of the complete correction ($r$=∞) for which $\mathrm{H}^{(\mathbf{k})} = \Xi^{(\mathbf{k})}$, and the case of no-correction ($r$=0) considered earlier, where

$$\tilde{\Xi}^{(\mathbf{k})} = \Xi^{(\mathbf{k})}.$$

The predistorted coefficients $\tilde{\Xi}^{(\mathbf{k})}$ depend on the required modal composition $\Xi^{(\mathbf{k})}$ and on the matrix elements of perturbations, Eq.(4.7).

When using the coefficients $\tilde{\Xi}^{(\mathbf{k})}$, modans are synthesized with the help of a predistorted function $\tilde{t}(\mathbf{x})$ instead of $t(\mathbf{x})$, where the beam $\tilde{t}(\mathbf{x})$ is determined by the equation similar to Eq.(2.7) with

$$\tilde{\mu}^{(\mathbf{k})} = \sqrt{\tilde{\kappa}^{(\mathbf{k})}} \exp\left[i\tilde{\alpha}^{(\mathbf{k})}\right], \quad \tilde{\alpha}^{(\mathbf{k})} = \alpha^{(\mathbf{k})} + \Delta\alpha^{(\mathbf{k})} \qquad (4.35)$$

The constant $c$, Eq.(2.33), is replaced by $\tilde{c}$ , respectively

$$\tilde{c}^{-1} = \max_{\mathbf{x}\in G}\left|\frac{\tilde{t}(\mathbf{x})}{w_0(\mathbf{x})/E_0}\right|$$

Let the choice of the predistorted parameters $\tilde{\kappa}^{(\mathbf{k})}$ and $\tilde{\alpha}^{(\mathbf{k})}$, which are near to the required values $\kappa^{(\mathbf{k})}$ and $\alpha^{(\mathbf{k})}$, be free for some time. Meanwhile in further consideration (see sections 4.4 and 4.5) we shall choose the parameters to ensure that the beam reconstructed from the perturbed modan provide both the required distribution $\kappa^{(\mathbf{k})}$ of the power of useful modes between the branches and the proper initial phase of the mode composition vector in each branch.

## 4.3 PERTURBATION MATRIX EVALUATION IN THE CASE OF SAMPLING THE MODAL FUNCTIONS

Phase sampling and quantization are important features that are crucial for manufacturing of computer generated optical elements as shown in the works [Huang, 1971] [Golub, 1990e]. The capabilities of digital holography are usually estimated in terms of comparison of the value of the hologram carrier spatial frequency with the spatial resolution of a mask-generating device. However, the recommendations based on such a method involve certain arbitrariness in judging diffraction order as fully separated and they really have empirical nature. Besides, such an important class of optical elements without the carrier spatial frequency as kinoforms goes beyond the scope of this method. Much more consistent approach is presented here that is based on studying the perturbations and errors of the beams formed by computer generated optical elements.

**A. Generalized model of quantization and sampling for computer generated optical elements**

All the fundamental stages of computer-aided synthesis of phase diffractive optical elements are, in one way or another, related to the phase quantization and sampling. A smooth phase φ is

(1) computed in the form of discrete pixels;

(2) reduced to the interval [0,2πm);

(3) quantized in to the given number of levels;

(4) used to control a photo mask plotter that images the phase in the form of variations of optical density on photo-plate that may be either gray-level or binary. Gray-level photo-masks can be used then to form a continuous-level diffractive microrelief using technology of photolithographic etching, gelatine or photopolymer technology. A set of binary masks allows us to obtain a staircase microrelief using a multi-step micro-lithographic technology.

Thus, the perturbations of sampling and sometimes quantization are unavoidable in principle for computer generated optical elements and must be estimated.

Note that quantization also takes place when holographic optical elements are produced using photolithography from interference photo-masks which are fabricated by physical holography techniques.

Let us formulate a model [Golub, 1990a] that is convenient to obtain the perturbation estimates and describes the sampling and quantization of optical elements manufactured with the help of a raster-scanning mask generator such as photo-plotter or raster e-beam machine.

The masks synthesized by means of a raster generator, consist of the set $J$ of separate "resolution cells" $G_{\mathbf{j}}$. The cells do not overlap. Their full set forms a total light aperture $G$ of a mask (see Figs. 4.2 and 4.3):

$$G = \bigcup_{\mathbf{j} \in J} G_{\mathbf{j}}, \quad G_{\mathbf{j}} \bigcap G_{\mathbf{j}'} = 0, \quad \mathbf{j} \neq \mathbf{j}',$$

where $\mathbf{j} = (j_1, j_2)$ is a 2-D multi-index number of a resolution cell centered at $\mathbf{x}_{\mathbf{j}}$. The size of the cells $G_{\mathbf{j}}$ determines the resolution of masks and those of optical elements replicated from the masks.

Define the uniform power distribution $\chi_{\mathbf{j}}(\mathbf{x})$ in a scanning spot $G_{\mathbf{j}}$ by specifying

$$\chi_{\mathbf{j}}(\mathbf{x}_{\mathbf{j}}) = 1; \quad \chi_{\mathbf{j}}(\mathbf{x}) = 0, \quad \mathbf{x} \notin G_{\mathbf{j}} \qquad (4.36)$$

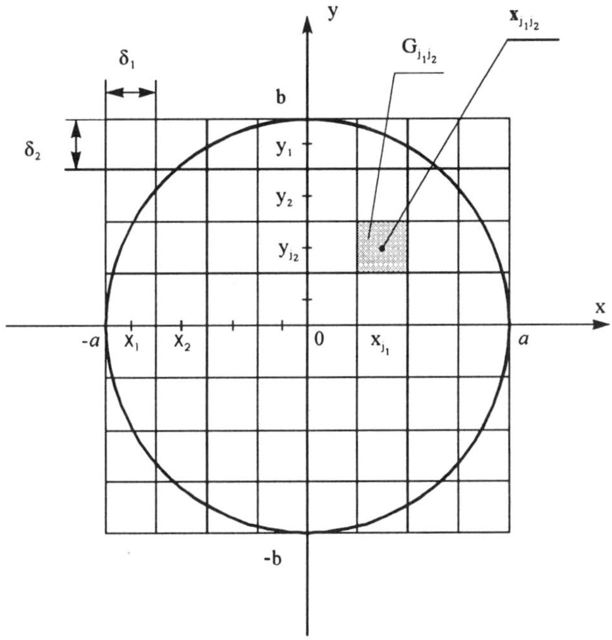


**Figure 4.2.** Mask structure generated by rectangular-raster writer (*2a, 2b* - mask sizes; $\delta_1$, $\delta_2$ - linear resolution along axis; $(x_{j_1}, y_{j_2})$ - coordinates of the center of the resolution cell number $(j_1, j_2)$, $G_{j_1,j_2}$ - resolution cell. (Modified from Golub M.A., Sisakyan I.N., Soifer V.A., Phase quantization and discretization in diffractive optics, *Proceedings SPIE, San-Diego'90*, 1334, 188, 1990).

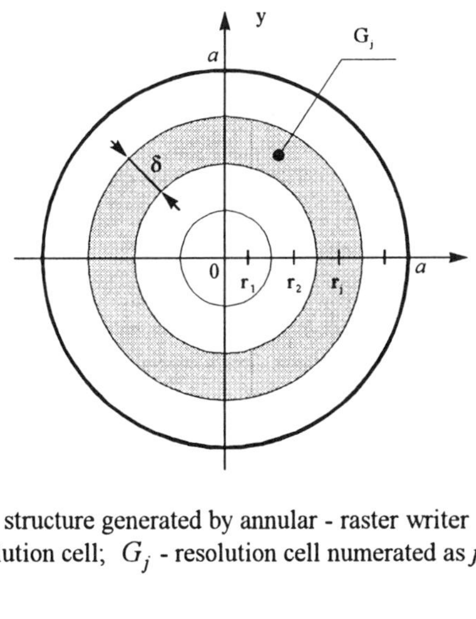


**Figure 4.3.** Mask structure generated by annular - raster writer (*2a* - diameter of the center of *j*-th resolution cell; $G_j$ - resolution cell numerated as *j*-th).

Raster-type mask-writers with rectangular (Fig 4.2) or annular (Fig 4.3) raster are most widely spread in technology equipment [Krackhardt, 1993]. Rectangular raster writers have feature-sizes of $\delta_1$ and $\delta_2$ that can be different along x and y axis respectively. Constant value $\hat{\Phi}(x_{j_1}, y_{j_2})$ is given to all the points within resolution cell $G_{j_1 j_2}$ This value is equal to the central pixel of phase function $\hat{\Phi}(x,y)$.

Annular-raster writers have ring-like resolution cell $\mathbf{G}_j$ with the width $\delta$ called radial resolution. A constant value $\hat{\Phi}(r_j)$ is given to all the points within j-th resolution cell $\mathbf{G}_j$. This value is equal to the pixel of function $\hat{\Phi}$ taken in the middle of the ring.

Thus, the sampling takes place of the phase φ and of the transfer function Γ, Eq.(2.34), The sampling can be specified mathematically by a piecewise approximation described by the linear sampling operator $\hat{D}$ as follows (see Fig. 4.4)

$$\hat{D}\Gamma(\mathbf{x}) = \sum_{\mathrm{j} \in J} \Gamma(\mathbf{x}_\mathrm{j})\chi_\mathrm{j}(\mathbf{x}), \quad \Gamma(\mathbf{x}_\mathrm{j}) = \exp\left[i\Phi_\mathrm{j}\right] \tag{4.37}$$

Procedure (4.37) implies replacement of the continuous transfer function $\Gamma(\mathbf{x})$, $\mathbf{x} \in G$, by the matrix of its pixels $\Gamma(\mathbf{x}_\mathrm{j})$ performed on the grid $\mathbf{x}_\mathrm{j} \in G$, with subsequent interpolation by the spot shaped as $\chi_\mathrm{j}(\mathbf{x})$.

Besides, the pixels of the phase $\Phi_\mathrm{j} = \mathrm{mod}_{2\pi m}\,\varphi(\mathbf{x}_\mathrm{j})$ which is brought to the interval $[0, 2\pi m]$, are quantized in $M$ equidistant levels with the step of $h_M = 2\pi m/M$ (see Fig. 4.5) thus being transformed to the pixels $\hat{\Phi}_\mathrm{j}$, $\mathrm{j} \in J$. The phase after quantization and sampling can be represented as

$$\hat{\Phi}(\mathbf{x}) = \sum_{\mathrm{j} \in J} \hat{\Phi}_\mathrm{j}\chi_\mathrm{j}(\mathbf{x})$$

With $\hat{\Phi}_\mathrm{j}$ taken as the control signals, the model of the quantized and sampled transfer function for the computer generated optical elements can be written as follows

$$\hat{\Gamma}(\mathbf{x}) = \sum_{\mathrm{j} \in J} \hat{\Gamma}_\mathrm{j}\chi_\mathrm{j}(\mathbf{x}); \quad \hat{\Gamma}_\mathrm{j} = \exp\left(i\hat{\Phi}_\mathrm{j}\right)$$

Following the last equation under the conditions (4.36) we can write down approximately

$$\hat{\Gamma}(\mathbf{x}) = \exp\left[i\hat{\Phi}(\mathbf{x})\right]$$

Let us discuss some particular cases. A uniform shift-invariant raster spot is being specified by the relationship

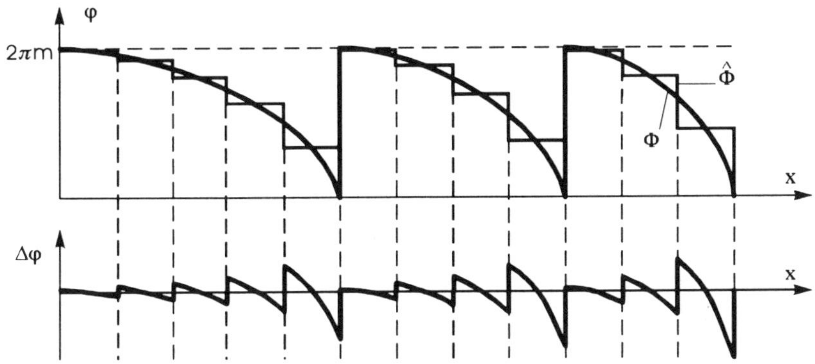


**Figure 4.4.** Phase sampling by piece-wise approximation. $\mathbf{\Phi}$ - piecewise continuous function; $\hat{\mathbf{\Phi}}$ - sampled version of $\mathbf{\Phi}$; $\Delta\varphi$ - perturbation of phase; $2\pi m$ - maximum of phase.

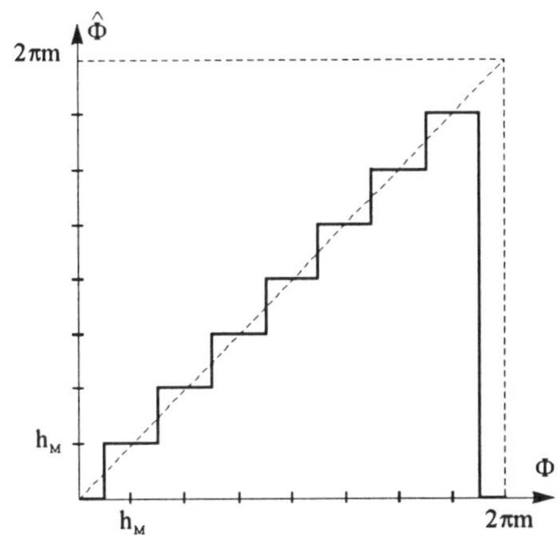


**Figure 4.5.** Nonlinearity, taking place under the uniform phase quantization. $\mathbf{\Phi}$ - phase value, $\hat{\mathbf{\Phi}}$ - quantized phase value, $h_M$ - quantization step, $2\pi m$ - maximum value of phase.

$$\chi_{\mathbf{j}}(\mathbf{x}) = \chi\left(\Omega_\delta\left(\mathbf{x} - \mathbf{x}_{\mathbf{j}}\right)\right)$$

and typical of the mask writers capable of appropriately correcting for the coordinates' non-linearity and of stabilizing the parameters, where $\chi(\cdot)$ is a spot function specified in normalized coordinates, $\chi(\mathbf{0}) = 1$, $\Omega_\delta$ is a matrix of spatial frequencies, with the eigenvalues being inverse to the sampling steps. In particular, the use of a rectangular raster mask writer with a uniform rectangular spot measured $\delta_1 \times \delta_2$ and having a size of $2a \times 2b$ (see Fig. 4.2) makes it possible to employ a specific model for which

$$G \subset [-a,a] \times [-b,b]; \;\; \mathbf{x_j} = (x_{j_1}, y_{j_2}); \;\; j_1 = \overline{1, N_1}, \;\; j_2 = \overline{1, N_2}$$

$$G_{\mathbf{j}} = \left[ x_{j_1} - \frac{\delta_1}{2}; \, x_{j_1} + \frac{\delta_1}{2} \right] \times \left[ y_{j_2} - \frac{\delta_2}{2}; \, y_{j_2} + \frac{\delta_2}{2} \right],$$

where the numbers $x_{j_1}$, $y_{j_2}$ produce an arithmetical progressions with the step of $\delta_1$ and $\delta_2$, respectively, $N_1\delta_1 = 2a$, $N_2\delta_2 = 2b$, and $\times$ denotes the direct product of the sets,

$$\chi(\mathbf{t}) = \text{rect}(t_1) \cdot \text{rect}(t_2), \;\; \mathbf{t} = (t_1, t_2) \tag{4.38}$$

$$\Omega_\delta = \begin{bmatrix} \delta_1^{-1} & 0 \\ 0 & \delta_2^{-1} \end{bmatrix}; \;\; \text{rect}(t_1) = \begin{cases} 1, & |t_1| < 0.5 \\ 0, & else \end{cases} \tag{4.39}$$

There are $N_1 \cdot N_2$ feasible positions of a spot within the frame $2a \times 2b$, with only its $|J|$ portion corresponding to the domain $G$.

When annular raster mask writers are used (see Fig. 4.3), the domain $G$ is a circle of the radius $a$, and the cells are numerated by a single index $j = \overline{1, N}$ and have the form of rings

$$\mathrm{G_j} = \left\{ \mathbf{x}: \; \mathrm{r_j} - \frac{\delta}{2} \le |\mathbf{x}| < \mathrm{r_j} + \frac{\delta}{2} \right\}$$

$$\chi_j(\mathbf{x}) \equiv \chi_j(r) = \text{rect}\left( \frac{r - r_j}{\delta} \right).$$

The numbers $r_j$ produce an arithmetical progression with the step $\delta$, $r = |\mathbf{x}|$. The annular mask writers are aimed at producing masks for radially symmetrical optical elements with transfer function

$$\Gamma = \Gamma(r); \;\; \Phi = \Phi(r) = \sum_{j=1}^{N} \Phi_j \chi_j(r)$$

**B. Approximation approach for estimating the matrix element**

Since we consider here the number of the quantization levels large enough, we can neglect the effects of quantization as compared with the effects of the sampling. It is in the case of sampling procedure (4.37) that the perturbed base-functions of Eq.(4.5) become piecewise functions with different values in different resolution cells $G_{\mathbf{j}}$, $\mathbf{j} \in J$ (see Fig. 4.1).

Let us build some convenient estimate [Golub, 1990b] for the matrix elements of the sampling-type perturbation. In assuming that for the slowly varying illuminating beam approximation takes place

$$w_0(\mathbf{x}) \cong w_0(\mathbf{x_j}), \quad \mathbf{x} \in G_\mathbf{j},$$

we find from Eqs.(4.3) and (4.5) that

$$\varphi_\mathbf{p}^{(\mathbf{k})}(\mathbf{x}) = \exp\left[-i2\pi\boldsymbol{\nu}(\mathbf{x}-\mathbf{x_j})\right]\psi_\mathbf{p}(\mathbf{x_j}), \quad \mathbf{x} \in G_\mathbf{j}, \quad \boldsymbol{\nu} = \boldsymbol{\nu}_\mathbf{n}^\mathbf{k}. \tag{4.40}$$

Since the values $\delta_1$ and $\delta_2$ are small, we can expand the mode function in each cell $G_\mathbf{j}$ into the power series up to the square terms of $\mathbf{x}$ included:

$$\psi_\mathbf{p}(\mathbf{x}) = \psi_\mathbf{p}(\mathbf{x_j}) + (\mathbf{x}-\mathbf{x_j})\nabla_\perp\psi_\mathbf{p}(\mathbf{x_j}) + \frac{1}{2}\left[(\mathbf{x}-\mathbf{x_j})\nabla_\perp\right]^2\psi_\mathbf{p}(\mathbf{x_j}), \quad \mathbf{x} \in G_\mathbf{j} \tag{4.41}$$

After substituting Eqs.(4.40) and (4.41) into Eq.(4.6), we obtain a representation of the perturbation function

$$\begin{aligned} h_\mathbf{p}^{(\mathbf{k})}(\mathbf{x}) &= \psi_\mathbf{p}(\mathbf{x_j})\left\{\exp\left[-i2\pi\boldsymbol{\nu}(\mathbf{x}-\mathbf{x_j})\right]-1\right\} - \\ &-\frac{1}{2}\left[(\mathbf{x}-\mathbf{x_j})\nabla_\perp\right]^2\psi_\mathbf{p}(\mathbf{x_j}), \quad \mathbf{x} \in G_\mathbf{j}, \quad \mathbf{j} \in J \end{aligned} \tag{4.42}$$

Equation (4.42) enables us to estimate the matrix elements of sampling-type perturbation in Eqs.(4.7) and (4.8) for $\rho_0(\mathbf{x}) \equiv 1$. Let us first estimate matrix elements

$$H_{\mathbf{pp}'}^{(\mathbf{k})} = \int_G h_\mathbf{p}^{(\mathbf{k})*}(\mathbf{x})\,\psi_{\mathbf{p}'}(\mathbf{x})d^2\mathbf{x} = \sum_{\mathbf{j}\in J}\int_{G_\mathbf{j}} h_\mathbf{p}^{(\mathbf{k})*}(\mathbf{x})\,\psi_{\mathbf{p}'}(\mathbf{x})d^2\mathbf{x} \tag{4.43}$$

We substitute Eqs.(4.42) and (4.41) into Eq.(4.43) and take into account the integration following equations:

$$\int_{G_\mathbf{j}} (\mathbf{x}-\mathbf{x_j})\, d^2\mathbf{x} = 0 \tag{4.44}$$

$$\int_{G_\mathbf{j}} (x-x_{j_1})\, d^2\mathbf{x} = \frac{\delta_1^3\delta_2}{12}; \quad \int_{G_\mathbf{j}} (y-y_{j_2})\, d^2\mathbf{x} = \frac{\delta_1\delta_2^3}{12} \tag{4.45}$$

$$\int_{G_\mathbf{j}} \left\{\exp\left[i2\pi\boldsymbol{\nu}(\mathbf{x}-\mathbf{x_j})\right]-1\right\} d^2\mathbf{x} = \delta_1\delta_2 C_1(\boldsymbol{\nu}) \tag{4.46}$$

$$\int_{G_\mathbf{j}} (\mathbf{x}-\mathbf{x_j})\left\{\exp\left[i2\pi\boldsymbol{\nu}(\mathbf{x}-\mathbf{x_j})\right]-1\right\} d^2\mathbf{x} = i\delta_1\delta_2\Omega_\delta^{-1}\mathbf{C}(\boldsymbol{\nu}), \tag{4.47}$$

with the following designations introduced:

$$\nabla_\delta = \Omega_\delta^{-1}\nabla_\perp = \left(\delta_1\frac{\partial}{\partial x}, \delta_2\frac{\partial}{\partial y}\right), \quad \Omega_\delta^{-1} = \begin{pmatrix}\delta_1 & 0 \\ 0 & \delta_2\end{pmatrix}, \quad \mathbf{c}(\mathbf{v}) = \begin{pmatrix} C_0(N_{\nu_1}, N_{\nu_2}) \\ C_0(N_{\nu_2}, N_{\nu_1}) \end{pmatrix} \tag{4.48}$$

$$C_0\left(N_{\nu_1}, N_{\nu_2}\right) = \frac{\operatorname{sinc}\left(\dfrac{1}{N_{\nu_1}}\right) - \cos\left(\dfrac{\pi}{N_{\nu_2}}\right)}{2\pi/N_{\nu_1}} \operatorname{sinc}\left(\frac{1}{N_{\nu_2}}\right) \tag{4.49}$$

$$c_1(\mathbf{v}) = 1 - \operatorname{sinc}\left(\frac{1}{N_{\nu_1}}\right)\operatorname{sinc}\left(\frac{1}{N_{\nu_2}}\right); \quad N_{\nu_1} = \frac{1}{\nu_x \delta_1}; \quad N_{\nu_2} = \frac{1}{\nu_y \delta_2} \tag{4.50}$$

The values $N_{\nu_1}$ and $N_{\nu_2}$ indicate the number of resolution cells $\delta_1$ and $\delta_2$ that fall on a period of carrier along $x$- and $y$- axis, respectively.

We found from Eqs.(4.43) - (4.50) that

$$\begin{aligned} H_{\mathbf{pp}'}^{(\mathbf{k})} &= \delta_1\delta_2 \sum_{\mathbf{j}\in J}\left\{-\frac{1}{12}\nabla_\delta \psi_{\mathbf{p}}^*(\mathbf{x_j})\nabla_\delta \psi_{\mathbf{p}'}(\mathbf{x_j}) - c_1(\boldsymbol{\nu})\psi_{\mathbf{p}}^*(\mathbf{x_j})\psi_{\mathbf{p}'}(\mathbf{x_j}) + \right. \\ &\left. + i\,\mathbf{c}(\boldsymbol{\nu})\psi_{\mathbf{p}}^*(\mathbf{x_j})\nabla_\delta \psi_{\mathbf{p}}(\mathbf{x_j})\right\} - \frac{1}{24}\left[\nabla_\delta^2 \psi_{\mathbf{p}}^*(\mathbf{x_j})\right]\psi_{\mathbf{p}'}(\mathbf{x_j}) \end{aligned} \tag{4.51}$$

Approximating the integral sum of Eq.(4.51) by the integral and using the orthonormality of the $\psi_{\mathbf{p}}$ functions, we obtain

$$\begin{aligned} H_{\mathbf{pp}'}^{(\mathbf{k})} &= -\frac{1}{12}\int_G \nabla_\delta \psi_{\mathbf{p}}^*(\mathbf{x})\nabla_\delta \psi_{\mathbf{p}'}(\mathbf{x})\,d^2\mathbf{x} - \delta_{\mathbf{pp}'} c_1\left(\boldsymbol{\nu}_{\mathbf{n}}^{(\mathbf{k})}\right) + \\ &+ i\,\mathbf{c}\left(\boldsymbol{\nu}_{\mathbf{n}}^{(\mathbf{k})}\right)\int_G \psi_{\mathbf{p}}^*(\mathbf{x})\nabla_\delta \psi_{\mathbf{p}}(\mathbf{x})\,d^2\mathbf{x} - \frac{1}{24}\int_G \left[\nabla_\delta^2 \psi_{\mathbf{p}}^*(\mathbf{x})\right]\psi_{\mathbf{p}'}(\mathbf{x})\,d^2\mathbf{x} \end{aligned} \tag{4.52}$$

To calculate the matrix elements $Q_{\mathbf{pp}'}^{(\mathbf{k},\mathbf{k}')}$ of Eqs.(4.8), we represent them as follows

$$Q_{\mathbf{pp}'}^{(\mathbf{k},\mathbf{k}')} = \sum_{\mathbf{j}\in J}\int_{G_{\mathbf{j}}} h_{\mathbf{p}}^{(\mathbf{k})*}(\mathbf{x})\,h_{\mathbf{p}'}^{(\mathbf{k}')}(\mathbf{x})\,d^2\mathbf{x}$$

From the expansion (4.42) with linear terms only and designations $\boldsymbol{\nu} = \boldsymbol{\nu}_{\mathbf{n}}^{(\mathbf{k})}$, $\boldsymbol{\nu}' = \boldsymbol{\nu}_{\mathbf{n}}^{(\mathbf{k}')}$ we obtain

$$\begin{aligned} Q_{\mathbf{pp}'}^{(\mathbf{k},\mathbf{k}')} &= \sum_{\mathbf{j}\in J}\int_{G_j}\left\{\psi_{\mathbf{p}}^*(\mathbf{x_j})\left[\exp\left(i2\pi\boldsymbol{\nu}(\mathbf{x}-\mathbf{x_j})\right)-1\right] - (\mathbf{x}-\mathbf{x_j})\nabla_\perp \psi_{\mathbf{p}}^*(\mathbf{x_j})\right\} \\ &\left\{\psi_{\mathbf{p}'}(\mathbf{x_j})\left[\exp\left(-i2\pi\boldsymbol{\nu}'(\mathbf{x}-\mathbf{x_j})\right)-1\right] - (\mathbf{x}-\mathbf{x_j})\nabla_\perp \psi_{\mathbf{p}'}^*(\mathbf{x_j})\right\}d^2\mathbf{x} \end{aligned}$$

The use of the standard integrals of Eqs.(4.44), (4.45), (4.48), and (4.47) and the final substitution of the integral sum taken over **j** by the integral taken over **x**, give us

$$Q_{\mathbf{pp}'}^{(\mathbf{k},\mathbf{k}')} = \delta_{\mathbf{pp}'}\left[c_1\left(\boldsymbol{\nu}_{\mathbf{n}}^{(\mathbf{k})}\right)+c_1\left(\boldsymbol{\nu}_{\mathbf{n}}^{(\mathbf{k}')}\right)-c_1\left(\boldsymbol{\nu}_{\mathbf{n}}^{(\mathbf{k}')}-\boldsymbol{\nu}_{\mathbf{n}}^{(\mathbf{k})}\right)\right]+$$
$$+\frac{1}{12}\int_G \nabla_\delta\psi_{\mathbf{p}}^*(\mathbf{x})\,\nabla_\delta\psi_{\mathbf{p}'}(\mathbf{x})\,d^2\mathbf{x}- \tag{4.53}$$
$$-i\,\mathbf{c}\left(\boldsymbol{\nu}_{\mathbf{n}}^{(\mathbf{k})}\right)\int_G \psi_{\mathbf{p}}^*(\mathbf{x})\nabla_\delta\psi_{\mathbf{p}'}(\mathbf{x})\,d^2\mathbf{x}+i\,\mathbf{c}\left(\boldsymbol{\nu}_{\mathbf{n}}^{(\mathbf{k}')}\right)\int_G \psi_{\mathbf{p}}^*(\mathbf{x})\nabla_\delta\psi_{\mathbf{p}'}(\mathbf{x})\,d^2\mathbf{x}$$

In particular, for real-valued mode functions we have

$$Q_{\mathbf{pp}'}^{(\mathbf{k},\mathbf{k})} \equiv \left\|h_{\mathbf{p}}^{(\mathbf{k})}\right\|^2 = \frac{1}{12}\int_G \left|\nabla_\delta\psi_{\mathbf{p}}(\mathbf{x})\right|^2 d^2\mathbf{x}+2c_1\left(\mathbf{v}_{\mathbf{n}}^{(\mathbf{k})}\right) \tag{4.54}$$

Perturbation matrix elements for the sampling are estimated by equations (4.52) and (4.53) with the accuracy as good as up to the $\delta^2$ order terms inclusive.

As an example, let us estimate the sampling errors in application to the Gauss-Hermite modes produced by the modan with the carrier $\boldsymbol{\nu}=(\nu_x,\nu_y)$. A beam waist is assumed to be located in the modan's plane and a plane wave illumination is assumed. In this case according to Eqs.(1.108) and (1.74) we can represent the derivatives by the expressions

$$\psi_p' = \frac{1}{\sigma}\left[\sqrt{p}\,\psi_{p-1}(x)+\sqrt{p+1}\,\psi_{p+1}(x)\right]$$
$$\psi_p'' = \frac{1}{\sigma^2}\left[\sqrt{(p-1)p}\,\psi_{p-2}(x)+(2p+1)\,\psi_p(x)+\sqrt{(p+1)(p+2)}\,\psi_{p+2}(x)\right] \tag{4.55}$$

following from the recurrence equations of the Hermite polynomials $H_p$, see [Janke, 1960]. Equation (4.55) and the property of the mode orthonormality, Eq.(2.5), make it possible to find simple expressions for the perturbation matrix elements. In the case of a plane illuminating wave ($w_0(\mathbf{x})=\text{const}$) and a square-shaped scanning spot sized $\delta=\delta_1=\delta_2$, we find from Eq.(4.52) the analytical formula

$$H_{plp'l'}^{(\mathbf{k})} = H_{plpl}^{(\mathbf{k})}\cdot\delta_{pp'}\delta_{ll'}+\frac{1}{24N_\sigma^2}\left\{\delta_{ll'}\left[\sqrt{(p-1)p}\,\delta_{p-2,p'}+\sqrt{(p+1)(p+2)}\,\delta_{p+2,p'}\right]+\right.$$
$$\left.+\delta_{pp'}\left[\sqrt{(l-1)l}\,\delta_{l-2,l'}+\sqrt{(l+1)(l+2)}\,\delta_{l+2,l'}\right]\right\}+$$
$$+i\frac{1}{N_\sigma}\left\{C_0\left(N_{\nu_1},N_{\nu_2}\right)\delta_{ll'}\left[\sqrt{p+1}\,\delta_{p+1,p'}-\sqrt{p}\,\delta_{p-1,p'}\right]+\right.$$
$$\left.+C_0\left(N_{\nu_1},N_{\nu_2}\right)\delta_{pp'}\left[\sqrt{l+1}\,\delta_{l+1,l'}-\sqrt{l}\,\delta_{l-1,l'}\right]\right\} \tag{4.56}$$

$$H_{plpl}^{(\mathbf{k})}=-\left\{c_1\left(\nu_{\mathbf{n}}^{(\mathbf{k})}\right)+\frac{p+l+1}{12N_\sigma^2}\right\} \tag{4.57}$$

where

$$N_\sigma=\sigma/\delta \tag{4.58}$$

Equation (4.54) gives us simple analytical formula

$$Q_{plpl}^{(\mathbf{k},\mathbf{k})}=2c_1\left(\mathbf{v}_{\mathbf{n}}^{(\mathbf{k})}\right)+\frac{p+l+1}{6N_\sigma^2} \tag{4.59}$$

Thus,

$$H_{plpl}^{(\mathbf{k})}=-\frac{1}{2}Q_{plpl}^{(\mathbf{k},\mathbf{k})} \tag{4.60}$$

It should be noted that all the matrix elements Eqs.(4.56)-(4.59) are the functions of the sum of indices *(p+l)*.

Note that Eqs.(4.56) through (4.60) contain only terms up to $\sim\delta^2$ order inclusive. We can see from Eq.(4.56) that excitation of the mode $\psi_{\mathbf{p}}$ with the aid of a sampled modan yields excitation of the parasitic modes numbered $(p',l')$, as shown in Fig. 4.6. The modes numbered as $p'=p\pm1$ or $l'=l\pm1$ have the phase shift of $\pi/2$ with respect to the mode $(p, l)$ and they disappear in the case when no carrier frequency **ν** exists.

If we took into account the terms higher than $\frac{1}{N_\sigma^2}$-, $\frac{1}{N_{\nu_1}^2}$-, and $\frac{1}{N_{\nu_2}^2}$-orders infinitesimal, we could estimate the power of the parasitic modes with the numbers $p'>p+2$, and $l'>l+2$. Being taken separately, these modes have minor powers each but they contribute considerably to the total power of parasitic modes $\Delta E_{\mathbf{p}}$.

Owing to Eqs.(4.13) and (4.60), we find here that

$$\Delta E_{\mathbf{p}}=Q-\frac{1}{4}Q^2$$

$$E_{\mathbf{p}}=\left(1-\frac{Q}{2}\right)^2,\quad \Delta E_{\mathbf{p}}+E_{\mathbf{p}}=1,$$

where $Q\equiv Q_{plpl}^{(\mathbf{k},\mathbf{k})}$.

It can be seen that the $\Delta E_{\mathbf{p}}, E_{\mathbf{p}}$ values are the functions of *(p+l)* that can treated as the mode order.

As is seen from Table 4.1 and Figs. 4.7 and 4.8, the total power $\Delta E_{\mathbf{p}}$ of parasitic modes can only amount to a fraction of a per cent or several per cents of the total beam power, provided that $N_\sigma$, $N_\nu\equiv N_{\nu_1}$ are large enough $\left(1/N_{\nu_2}=0\right)$.

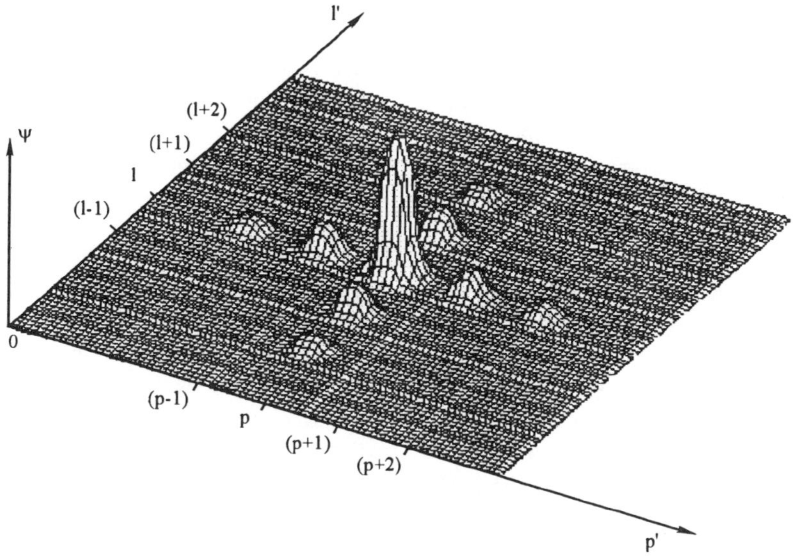


**Figure 4.6.** Numbers of the parasitic modes that are excited upon sampling. *(p,l)*- required mode number.

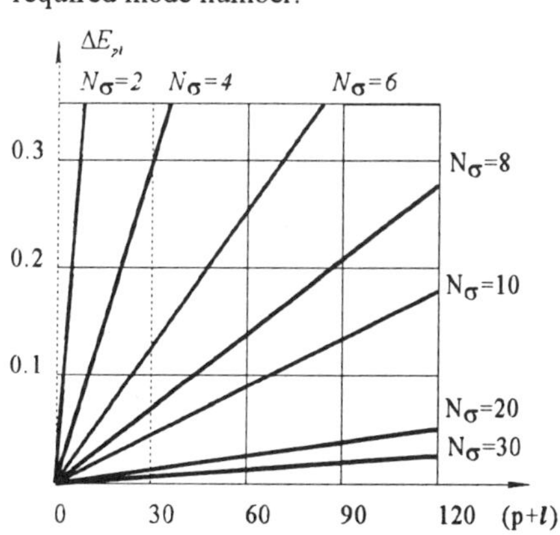


**Figure 4.7.** A total power $\Delta E_{pl}$ of the parasitic modes upon exciting the *(p, l)* mode with the aid of a sampled modan without carrier $(1/N_{\nu}=0)$ *(p+l)* - order of the required mode.

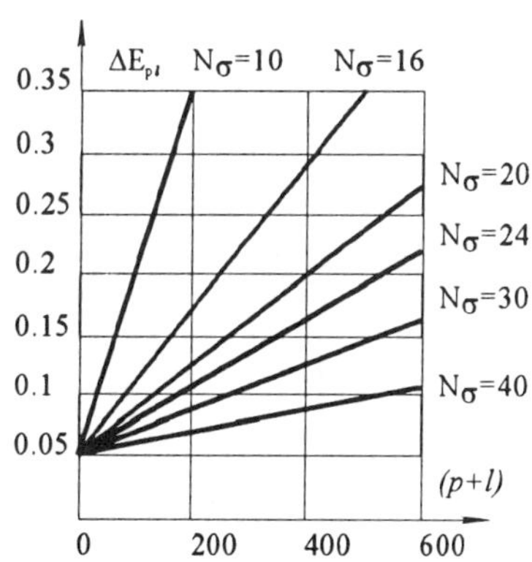


**Figure 4.8.** A total power $\Delta E_{pl}$ of the parasitic modes for the modan with the carrier $N_{\nu}=8$; *(p+l)* - order of the required mode.

**Table 4.1**
**Values of $\Delta E_{pl}$, $p+l=10$**

| $N_{\sigma}$ | $N_{\nu}$: | **4** | **6** | **8** | **10** | **20** | **50** |
|---|---|---|---|---|---|---|---|
| 10 | | 0.205 | 0.105 | 0.068 | 0.050 | 0.026 | 0.020 |
| 100 | | 0.189 | 0.088 | 0.050 | 0.033 | 0.008 | 0.001 |

## 4.4. ENERGY EFFICIENCY OF MODANS

The energy efficiency is defined here as the portion of the full power of illuminating beam that is directed to the useful beams. A common-used term "diffraction efficiency" appears to be somewhat irrelevant when operating with an axial beam, that is, with zero diffraction order. "Nevertheless" diffraction efficiency" and "energy efficiency" mean the same for the holographic elements working in the first (or higher) diffraction orders.

The energy efficiency of a thin optical element (upon the phase quantization, wavelength changed, and the maximum phase mismatching) is commonly estimated by means of expansion in terms of diffraction orders. A decrease in a modan's energy efficiency is caused by (1) diffraction coding of the complex transfer function into the phase one (see section 2.5), (2) sampling over arguments and perhaps quantization over the levels of the coded phase (see section 4.1). This section is aimed at obtaining simple approximate estimates of the modan energy efficiency depending on: a) the modal composition of a formed beam; b) the method of coding; c) the modan resolution; and d) the number of the pixel for phase sampling.

We call as the modan's energy efficiency the portion $\hat{\varepsilon}_{\mathbf{n}}$ of the illuminating beam power that is spent on the excitation of the required modes $\psi_{\mathbf{p}}$, $\mathbf{p} \in L^{(\mathbf{k})}$, $\mathbf{k} \in K$ in all $L$ branches within the $\mathbf{n}$-th working diffraction order.

Let us first consider the simple case of modan without any perturbations. For the transfer function, which is phase-only and obtained upon coding, the total power of the $\mathbf{n}$-th diffraction order $W_{\mathbf{n}}$, Eq.(2.46), goes on the required modes $\psi_{\mathbf{p}}$, $\mathbf{p} \in L^{(\mathbf{k})}$, $\mathbf{k} \in K$, and we can find the energy efficiency, designated as $\varepsilon_{\mathbf{n}}$ from the equation

$$\varepsilon_{\mathbf{n}} = \frac{\|W_{\mathbf{n}}\|^2}{E_0}, \quad E_0 = \|w_0\|^2. \tag{4.61}$$

Equation (2.45) and (4.61) give us

$$\varepsilon_{\mathbf{n}} = g_{\mathbf{n}}^2 c^2 \tag{4.62}$$

The values

$$\varepsilon_{\mathbf{n}}^{(\mathbf{k})} \equiv \frac{\left\|\mathrm{W}_{\mathbf{n}}^{(\mathbf{k})}\right\|^2}{E_0} = \kappa^{(\mathbf{k})}\varepsilon_{\mathbf{n}}, \qquad \sum_{\mathbf{k}\in\mathrm{K}} \varepsilon_{\mathbf{n}}^{(\mathbf{k})} = \varepsilon_{\mathbf{n}} \tag{4.63}$$

are the energy efficiencies of the branches. The value $\varepsilon_{\mathbf{n}}$ depends on: (a) the value $g_{\mathbf{n}}$, Eq.(2.39), (b) the method of coding (section 2.5), and (c) the constant $c$, Eq.(2.33) that describes the procedure of reducing the amplitude transmittance of the calculated modan's transfer function (2.32) $T$ to the [0,1] interval. The coefficient $g_{\mathbf{n}}^2$ is the diffraction efficiency of the basic diffraction grating in the **n**-th diffraction order and does not depend on $T$ under the condition (2.4). The value $1\text{-}g_{\mathbf{n}}^2$ characterizes the energy losses resulted from the fact that a portion of the illuminating beam power $w_0$ is directed to non-operative diffraction orders. It is noteworthy that the $g_{\mathbf{n}}$ coefficient essentially depends on the number of the working diffraction order and it can be equal 1 in zero (axial) working order (see table 2.3). Equation (4.62) shows that the energy efficiency of the coding procedure is proportional to $g_{\mathbf{n}}^2$. Thus we can conclude that the energy efficiencies of various methods for coding optical elements relate to each other the same manner as the diffraction efficiencies of corresponding basic diffraction gratings in their working orders. In particular, phase optical elements can ensure the same improvement in the energy efficiency, compared with amplitude elements, as a phase sinusoidal diffraction grating compared with an amplitude grating. Of course, beam-separation condition (2.4) must be obeyed to ensure such conclusions.

Furthermore, the phase elements can provide an additional improvement in energy efficiency by factor $\left[g_{(0,0)}/g_{(1,0)}\right]^2$ if we change from the first working order to the zero order.

Owing to Eq.(2.33), the $c$ constant is in inverse proportion to the ratio of the output beam amplitude to the input beam amplitude, provided that they are normalized to the same power. If the amplitudes' ratio shows the sharp maximum, the $c$ constant is small, with the energy efficiency being low. In opposite under the condition that the beams normalized to the same power have close-varying amplitudes, the ratio of their amplitudes varies smoothly around an average value, with the $c$ constant being near to the unity. Last case enables us to achieve high energy efficiency in the course of diffractive coding of an optical element. If the calculated transfer function is phase-only, the ratio of the beam amplitudes equals one, in this case $c$=1, and the natural conclusion will be that phase-only optical elements have the limiting energy efficiency equal to one. Therefore, we can characterize the $c$ value as the coefficient of the amplitude uniformity.

Now we can consider the case when perturbations of sampling are present and correction is performed. The power

$$\left\|\hat{W}_{\mathbf{n}}^{(\mathbf{k})}\right\|^2 = E_0 \varepsilon_{\mathbf{n}}^{(\mathbf{k})} \cdot \frac{\left\|\mathrm{H}^{(\mathbf{k})}\right\|^2}{\left|\tilde{\Xi}^{(\mathbf{k})}\right|^2} \tag{4.64}$$

of the beam (see Eqs.(4.3), (4.22), (4.28), which is directed to the **k**-th branch of modan, is separated into the powers of the required modes, Eq.(4.29), and the power of the orthogonal parasitic modes, Eq.(4.30). In accordance with Eq.(4.29) and Eq.(4.64), we can find the portion $\hat{\varepsilon}_{\mathbf{n}}^{(\mathbf{k})}$ of the power that accounts for useful modes of **k**-th branch as

$$\hat{\varepsilon}_{\mathbf{n}}^{(\mathbf{k})} = \varepsilon_{\mathbf{n}}^{(\mathbf{k})} \cdot \frac{\left|\mathrm{H}^{(\mathbf{k})}\right|^2}{\left|\tilde{\Xi}^{(\mathbf{k})}\right|^2}. \tag{4.65}$$

In this case the power distribution between branches differs from the desired distribution $\{\kappa^{(\mathbf{k})}\}$.

Even in the presence of the perturbations we can still achieve the power distribution of useful modes between branches that agrees with the required distribution. This can be done if we introduce into the beam $t(\mathbf{x})$, Eqs.(2.2), (2.3), predistorted numbers $\tilde{\kappa}^{(\mathbf{k})}$ instead of $\kappa^{(\mathbf{k})}$ and choose $\tilde{\kappa}^{(\mathbf{k})}$ so that $\varepsilon_{\mathbf{n}}^{(\mathbf{k})}$ should be distributed in the same manner as $\hat{\varepsilon}_{\mathbf{n}}^{(\mathbf{k})}$. Thus we can write down

$$\hat{\varepsilon}_{\mathbf{n}}^{(\mathbf{k})} = \tilde{q}\varepsilon_{\mathbf{n}}^{(\mathbf{k})},$$

where $\tilde{q}$ is the coefficient of proportionality, which is independent of **k**. We find then from Eqs.(4.65), (4.63), and (4.62) the equation

$$\hat{\varepsilon}_{\mathbf{n}}^{(\mathbf{k})} = g_{\mathbf{n}}^2 \tilde{c}^2 \tilde{\kappa}^{(\mathbf{k})} \cdot \frac{\left|\mathrm{H}^{(\mathbf{k})}\right|^2}{\left|\tilde{\Xi}^{(\mathbf{k})}\right|^2} = \tilde{q} g_{\mathbf{n}}^2 c^2 \kappa^{(\mathbf{k})},$$

that can be solved with respect to $\tilde{\mathrm{H}}^{(\mathbf{k})}$ as follows:

$$\tilde{\kappa}^{(\mathbf{k})} = \tilde{q}\kappa^{(\mathbf{k})} \cdot \frac{\left|\tilde{\Xi}^{(\mathbf{k})}\right|^2}{\left|\mathrm{H}^{(\mathbf{k})}\right|^2}. \tag{4.66}$$

The condition of normalizing

$$\sum_{\mathbf{k}} \tilde{\kappa}^{(\mathbf{k})} = 1$$

enables us to determine

$$\tilde{q} = \left(\frac{\tilde{c}}{c}\right)^2 q, \tag{4.67}$$

where

$$\frac{\tilde{c}}{c} = \frac{\max\limits_{\mathbf{x}\in G}|t(\mathbf{x})|}{\max\limits_{\mathbf{x}\in G}|\tilde{t}(\mathbf{x})|} \tag{4.68}$$

is the coefficient characterizing decrease in the energy efficiency caused by the changing variation range of $|t|$, and q is a new coefficient defined by formula

$$q = \left[ \sum_{\mathbf{k}\in K} \kappa^{(\mathbf{k})} \cdot \frac{|\tilde{\Xi}^{(\mathbf{k})}|^2}{|\mathrm{H}^{(\mathbf{k})}|^2} \right]^{-1} \tag{4.69}$$

The value q is the coefficient of the decrease in the energy efficiency of multibeam modan due to perturbations resulted from the predistortions $\tilde{\kappa}^{(\mathbf{k})}$ and $\tilde{\Xi}^{(\mathbf{k})}$ introduced into the beam $\tilde{t}$ .

On performing correction (4.31) and predistortions (4.66), the predistorted beam in the **k**-th branch, Eq.(4.3), can be written as

$$\hat{W}_{\mathbf{n}}^{(\mathbf{k})}(\mathbf{x}) = \sqrt{E_0 \varepsilon_{\mathbf{n}} q}\, \mu^{(\mathbf{k})} \exp\left[i\Delta\alpha^{(\mathbf{k})}\right] \frac{\eta^{(\mathbf{k})}(\mathbf{x})}{\left|\mathrm{H}^{(\mathbf{k})}\right|} \exp\left[i2\pi \boldsymbol{\nu}_{\mathbf{n}}^{(\mathbf{k})} \mathbf{x}\right] \tag{4.70}$$

It differs from the required multibeam beam $t(\mathbf{x})$, Eq.(2.2), by the factor $\sqrt{q}$ and by $\frac{\xi^{(\mathbf{k})}}{\left|\Xi^{(\mathbf{k})}\right|}$ replaced with $\frac{\eta^{(\mathbf{k})}}{\left|\mathrm{H}^{(\mathbf{k})}\right|}$.

Making use of Eqs.(4.33) and (4.31), we can write $q$ as

$$q = \left[ \sum_{\mathbf{k}\in K} \kappa^{(\mathbf{k})} \cdot \frac{\left(C^{(\mathbf{k})}\Xi^{(\mathbf{k})}, \Xi^{(\mathbf{k})}\right)}{\left(\left(E^{(\mathbf{k})} + R_r^{(\mathbf{k})}\right)\Xi^{(\mathbf{k})}, \Xi^{(\mathbf{k})}\right)} \right]^{-1} \tag{4.71}$$

and represent the value $\left|\mathrm{H}^{(\mathbf{k})}\right|^2$ as the Hermite non-negatively defined quadratic form with respect to $\tilde{\Xi}^{(\mathbf{k})}$ vector:

$$\left|\mathrm{H}^{(\mathbf{k})}\right|^2 = \left(\left[E^{(\mathbf{k})} + R_0^{(\mathbf{k})}\right]\tilde{\Xi}^{(\mathbf{k})}, \tilde{\Xi}^{(\mathbf{k})}\right) \tag{4.72}$$

where

$$C^{(\mathbf{k})} = \sum_{r'=0}^{r} \left[-H^{(\mathbf{k})}\right]^{r'} \sum_{r''=0}^{r} \left[-H^{(\mathbf{k})*}\right]^{r''}, \tag{4.73}$$

$$R_r^{(\mathbf{k})} = H_r^{(\mathbf{k})} + H_r^{(\mathbf{k})*} + H_r^{(\mathbf{k})} \cdot H_r^{(\mathbf{k})*} \tag{4.74}$$

$$R_0^{(\mathbf{k})} = H^{(\mathbf{k})} + H^{(\mathbf{k})*} + H^{(\mathbf{k})} \cdot H^{(\mathbf{k})*}. \tag{4.75}$$

It must be noted that the matrix $R_r^{(\mathbf{k})}$ becomes simply a value (4.21) in the case of single-modal modans.

It is well known that the values of the Hermite non-negatively defined quadratic form allow one to obtain estimates by means of eigenvalues of its matrix:

$$\left[1 + \lambda_{\min}\left(R_0^{(\mathbf{k})}\right)\right]\left|\tilde{\Xi}^{(\mathbf{k})}\right|^2 \le \left|\mathrm{H}^{(\mathbf{k})}\right|^2 \le \left[1 + \lambda_{\max}\left(R_0^{(\mathbf{k})}\right)\right]\left|\tilde{\Xi}^{(\mathbf{k})}\right|^2, \tag{4.76}$$

where $\lambda_{\min}\left(R_0^{(\mathbf{k})}\right)$ and $\lambda_{\max}\left(R_0^{(\mathbf{k})}\right)$ are the minimum and maximum eigenvalues of the matrix $R_0^{(\mathbf{k})}$.

If the mode composition of the required modal beams is arbitrary, then according to Eqs.(4.69), (4.72), and (4.76), the inequality

$$\left[\sum_{\mathbf{k}\in K}\frac{\kappa^{(\mathbf{k})}}{1+\lambda_{\min}\left(R_0^{(\mathbf{k})}\right)}\right]^{-1}\le q\le\left[\sum_{\mathbf{k}\in K}\frac{\kappa^{(\mathbf{k})}}{1+\lambda_{\max}\left(R_0^{(\mathbf{k})}\right)}\right]^{-1} \tag{4.77}$$

determines the range of variations of the $q$ coefficient of the energy efficiency reduction.

Note that when the $\tilde{\Xi}^{(\mathbf{k})}$ vectors of mode composition are the eigenvectors for the matrix $R_0^{(\mathbf{k})}$, Eq.(4.75), the boundaries of the estimate (4.77) are reached.

A single-modal modan obeys the condition $\lambda_{\max}(\cdot)=\lambda_{\min}(\cdot)$, and the coefficient of the energy efficiency reduction caused by perturbations can be found from the simple expression

$$q=\left[\sum_{\mathbf{p}\in K}\frac{\kappa^{(\mathbf{p})}}{\left|1+H^{(\mathbf{p})}\right|^2}\right]^{-1}. \tag{4.78}$$

***Example.***

Let a four-beam single-modal modan be illuminated by a plane wave and form four modes of Gauss-Hermite, Eq.(1.108), with the $(p,l)$ numbers of the same order, $(p+l)$=const. There is no correction, that is, $r$=0. A method is used of the phase coding with the spatial frequencies of branches having the form $(\nu,0)$, $(-\nu,0)$, $(0,\nu)$, and $(0,-\nu)$, where $\nu>0$ is an absolute value of the spatial frequency. Also, the $\sigma$ width of the principal mode Gaussian beam at the $\exp(-2)\cong 0.14$ level of the intensity fall is specified. Perturbations are caused by the sampling, that is, by the replacement of a continuous mode function by the matrix of its pixels taken on a rectangular grid with the same step of $\delta$ along $x$- and $y$-axes (see Fig. 4.1). The calculation of matrix elements using Eqs.(4.57) and (4.60) gives us the following result:

$$H^{(\mathbf{p})}\equiv H=-\frac{p+l+1}{6N_\sigma^2}-\left[1-\operatorname{sinc}\left(\frac{1}{N_\sigma}\right)\right] \tag{4.79}$$

$$Q^{(\mathbf{p})}\equiv Q=-2H, \tag{4.80}$$

where

$$N_\sigma=\frac{\sigma}{\delta},\quad N_\nu=\frac{1}{\nu\delta} \tag{4.81}$$

The first term in Eq.(4.79) describes the error of the discrete representation of the mode function $\psi_{pl}$, and the second term describes the error of discrete reproduction of the carrier spatial frequency.

The coefficient $q$ of the energy efficiency reduction is characterized here by the expression

$$q = (1+H)^2 . \tag{4.82}$$

For $\delta \to 0$, we have $N_\nu \to \infty$, $N_\sigma \to \infty$ and, as it should be expected, in accordance with Eq (4.78) $q \to 1$.

Table 4.2 and Fig. 4.9 show the values of $q$ depending on $N_\nu$, $N_\sigma$ and $p+l$. The energy efficiency is seen to fall with the increasing order $(p+l)$ of the modes $\psi_{\mathrm{pl}}$ and to increase with the increasing $N_\sigma$ or $N_\nu$.

When separate modes are formed from a plane wave, equations (4.62), (2.33), (1.96), and (1.108) give us estimation of the energy efficiency of the Gaussian modans:

$$\varepsilon_n \geq g_n^2 \frac{\pi}{2} \cdot \frac{\sigma^2}{|G|}$$

In particular, for the Gauss-Laguerre modans we have

$$\varepsilon_n \geq g_n^2 \frac{\sigma^2}{2a^2}$$

and for the Gauss-Hermite modans

$$\varepsilon_n \geq g_n^2 \frac{\pi \sigma^2}{8ab}$$

Right-hand sides of the estimates correspond to a principal mode, that is, to a Gaussian beam, which is limited by the region $G$ and formed from a plane beam. Upon forming the modes of higher orders $(p,l)$, the value $\varepsilon_n$ is determined by the corrected expressions

$$\varepsilon_n^{(p,l)} = t_{pl} g_n^2 \frac{\pi}{2} \cdot \frac{\sigma^2}{|G|},$$

where $t_{pl}$ is a coefficient that exceeds the unity and is found from numerical studies of maximums of the functions (1.108) and (1.96). For example, for the Gauss-Hermite modes we have $t_{pl} = t_p \cdot t_l$, with the values $t_p$ given in the table.

The factor $\sigma^2 / |G|$ characterizes the energy losses arising upon transformation of a plane wave into a Gaussian beam. These losses can be reduced by using the Gaussian illuminating beam with the parameter $\sigma_0 > \sigma$.

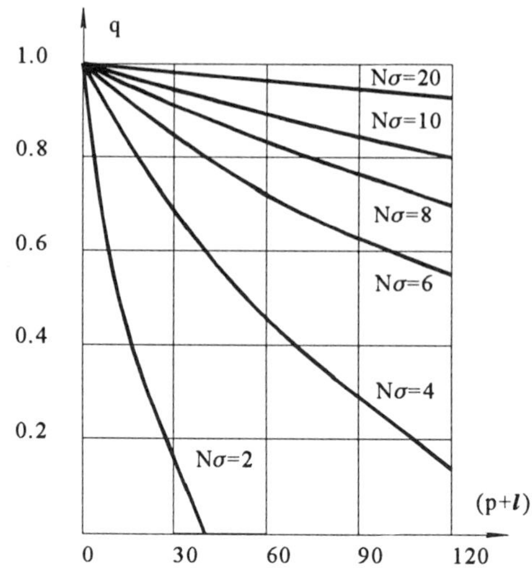


**Figure 4.9.** The coefficient $q$ of the energy efficiency reduction due to sampling the Gauss-Laguerre modans ($1/N_\nu = 0$); *(p+l)* - order of the required mode.

**Table 4.2**
**The values of the coefficient of the energy efficiency reduction *(p+l=10)***

| $N_\sigma$ $N_\nu$: | 4 | 6 | 8 | 10 | 20 | 50 |
|---|---|---|---|---|---|---|
| 10 | 0.794 | 0.895 | 0.932 | 0..95 | 0.974 | 0.980 |
| 100 | 0.810 | 0.912 | 0.950 | 0.976 | 0.992 | 0.999 |

Modified from Golub M.A., Sisakyan I.N., Soifer V.A., Computer generated optical elements for fiber's mode selection and launching, *Proceedings SPIE, OE/Fibers'90*, 1365, 156, 1990

**Table 4.3.**
**Corrective coefficient for energy efficiency of higher-order modans**

| p | 0 | 1 | 2 | 3 | 4 | 5 | 6 | 7 | 8 | 9 | 10 | 11 |
|---|---|---|---|---|---|---|---|---|---|---|---|---|
| $t_{p0}$ | 1 | 1.36 | 1.59 | 1.64 | 1.72 | 0.8 | 1.85 | 1.91 | 1.93 | 1.99 | 2.09 | 2.12 |

## 4.5. ACCURACY OF ANALYSIS OR FORMATION OF COHERENT BEAMS

It was already stated in section 4.3 that the computer-aided synthesis of phase modans always causes the excitation of parasitic modes in addition to required ones, in accordance with Eq.(4.10). To estimate the quality of modans for the tasks of forming the mode composition, one can make use of empirical considerations based on comparing the mode size $\sigma$ and the $1/|\nu_n|$ period of the carrier diffraction grating, with the step of sampling equal to $\delta$.

In this work the authors suggest a quantitative method to estimate modan's quality that provide proper relations between $1/|\nu_{\mathbf{n}}|$ and δ upon a certain acceptable modan's accuracy. The accuracy performances are introduced that offer a quantitative estimate of the difference between the required mode composition of a light beam and the one, which is really formed.

Let us compare the operation of a multibeam modan in the presence and absence of perturbations using the matrix elements of Eqs.(4.7) - (4.9) introduced in section 4.3. In the absence of perturbations the modan forms, in its **k**-th branch, a field $W_{\mathbf{n}}^{(\mathbf{k})}$, Eq.(2.47) that changes to the field $\hat{W}_{\mathbf{n}}^{(\mathbf{k})}$, Eq.(4.70), in the presence of perturbation. The perturbation effect yields the replacement in Eq.(2.47) of the beams $\xi^{(\mathbf{k})}$, Eq.(2.6), with the perturbed beams $\eta^{(\mathbf{k})}$, Eq.(4.4). The energy efficiency reduction that is taken into account as a factor $\sqrt{q}$ in Eq.(4.70) also takes place.

In order to estimate the accuracy in formation of the mode composition in the **k**-th branch, let us introduce the accuracy parameter

$$\Delta_f^{(\mathbf{k})} = \left\| \frac{1}{\sqrt{q}} \hat{W}_{\mathbf{n}}^{(\mathbf{k})} - W_{\mathbf{n}}^{(\mathbf{k})} \right\|. \tag{4.83}$$

The square of the value (4.83) is a power of the differential light field that arises when the powers $\hat{W}_{\mathbf{n}}^{(\mathbf{k})}$ and $W_{\mathbf{n}}^{(\mathbf{k})}$ are applied to the branches of a virtual twin-wave interferometer turned to the zero fringe with the path difference of π for the branches. The $1/\sqrt{q}$ normalization is introduced with the aim of reduction the scale of values of $\hat{W}_{\mathbf{n}}^{(\mathbf{k})}$ to the initial scale of values of $W_{\mathbf{n}}^{(\mathbf{k})}$. A total error in the formation of mode composition by means of multibeam modan can be designated as $\Delta_f^2$ and defined by the sum

$$\Delta_f^2 = \sum_{\mathbf{k}\in K} \left[\Delta_f^{(\mathbf{k})}\right]^2 = E_0 \varepsilon_{\mathbf{n}} \delta_f^2, \tag{4.84}$$

where

$$\delta_f^2 = \frac{\Delta_f^2}{\sum_{\mathbf{k}\in K} \left\| W_{\mathbf{n}}^{(\mathbf{k})} \right\|^2} = \sum_{\mathbf{k}\in K} \left[\delta_f^{(\mathbf{k})}\right]^2 = \delta_s^2 + \delta_d^2 \tag{4.85}$$

is a relative error of mode formation that can be found from Eqs.(4.83), (2.47), and (4.70) in the form

$$\left[\delta_f^{(\mathbf{k})}\right]^2 = \kappa^{(\mathbf{k})} \left\| \exp\left[i\Delta\alpha^{(\mathbf{k})}\right] \frac{\eta^{(\mathbf{k})}}{\left|\mathrm{H}^{(\mathbf{k})}\right|} - \frac{\xi^{(\mathbf{k})}}{\left|\Xi^{(\mathbf{k})}\right|} \right\|^2. \tag{4.86}$$

By virtue of Eqs.(4.22) and (4.23), the relative error in Eqs.(4.84) and (4.85) contains two components:

$$\left[\delta_f^{(\mathbf{k})}\right]^2 = \left[\delta_s^{(\mathbf{k})}\right]^2 + \left[\delta_d^{(\mathbf{k})}\right]^2.$$

The component

$$\left[\delta_s^{(\mathbf{k})}\right]^2 = \kappa^{(\mathbf{k})} \left| \exp\left[i\Delta\alpha^{(\mathbf{k})}\right] \frac{\mathrm{H}^{(\mathbf{k})}}{\left|\mathrm{H}^{(\mathbf{k})}\right|} - \frac{\Xi^{(\mathbf{k})}}{\left|\Xi^{(\mathbf{k})}\right|} \right|^2 \tag{4.87}$$

characterizes both the difference between the modal compositions

$$\frac{\mathrm{H}^{(\mathbf{k})}}{\left|\mathrm{H}^{(\mathbf{k})}\right|} \text{ and } \frac{\Xi^{(\mathbf{k})}}{\left|\Xi^{(\mathbf{k})}\right|}$$

with their adjustment to the same initial phase.

The component

$$\left[\delta_d^{(\mathbf{k})}\right]^2 = \kappa^{(\mathbf{k})} \frac{\left\|\Delta\eta^{(\mathbf{k})}\right\|^2}{\left|\mathrm{H}^{(\mathbf{k})}\right|^2} \tag{4.88}$$

characterizes the changes in a beam form caused by parasitic modes. To estimate $\delta_d^2$ let us take into account that Eqs.(4.28), (4.29), (4.4), (4.72), (4.6) - (4.9) give us

$$\left\|\Delta\eta^{(\mathbf{k})}\right\|^2 = \left\|\eta^{(\mathbf{k})}\right\|^2 - \left|\mathrm{H}^{(\mathbf{k})}\right|^2 = \sum_{\mathbf{p}\in L^{(\mathbf{k})}} \sum_{\mathbf{p}'\in L^{(\mathbf{k})}} \tilde{\xi}_{\mathbf{p}}^{(\mathbf{k})*} \tilde{\xi}_{\mathbf{p}'}^{(\mathbf{k})} \cdot$$

$$\left(\delta_{\mathbf{pp}'} + H_{\mathbf{pp}'}^{(\mathbf{k})} + H_{\mathbf{pp}'}^{(\mathbf{k})*} + Q_{\mathbf{pp}'}^{(\mathbf{k},\mathbf{k})}\right) - \left|\mathrm{H}^{(\mathbf{k})}\right|^2 = \left(\tilde{A}^{(\mathbf{k})}\tilde{\Xi}^{(\mathbf{k})}, \tilde{\Xi}^{(\mathbf{k})}\right)$$

where

$$\tilde{A}^{(\mathbf{k})} = Q^{(\mathbf{k})} - H^{(\mathbf{k})} H^{(\mathbf{k})*} \tag{4.89}$$

After substituting the expressions describing $\tilde{\Xi}^{(\mathbf{k})}$, Eq.(4.31), we obtain

$$\left\|\Delta\eta^{(\mathbf{k})}\right\|^2 = \left(A^{(\mathbf{k})}\Xi^{(\mathbf{k})}, \Xi^{(\mathbf{k})}\right), \tag{4.90}$$

where

$$A^{(\mathbf{k})} = \sum_{r'=0}^{r} \left(-H^{(\mathbf{k})}\right)^{r'} \tilde{A}^{(\mathbf{k})} \sum_{r'=0}^{r} \left(-H^{(\mathbf{k})*}\right)^{r'} \tag{4.91}$$

Another form that accounts of the expression

$$\tilde{\Xi}^{(\mathbf{k})} = \left(E^{(\mathbf{k})} + H^{(\mathbf{k})*}\right)^{-1} \mathrm{H}^{(\mathbf{k})},$$

derived from Eq.(4.33), gives us

$$\left\|\Delta\eta^{(\mathbf{k})}\right\|^2 = \left(B^{(\mathbf{k})}\mathrm{H}^{(\mathbf{k})}, \mathrm{H}^{(\mathbf{k})}\right), \tag{4.92}$$

where

$$B^{(\mathbf{k})} = \left(E^{(\mathbf{k})} + H^{(\mathbf{k})}\right)^{-1} \tilde{A}^{(\mathbf{k})}\left(E^{(\mathbf{k})} + H^{(\mathbf{k})*}\right)^{-1} \tag{4.93}$$

Thus the value of $\delta_d^2$ specified by Eqs.(4.88), (4.92) and (4.85) is written now in the final form

$$\delta_d^2 = \sum_{\mathbf{k}\in K} \kappa^{(\mathbf{k})} \frac{\left(A^{(\mathbf{k})}\Xi^{(\mathbf{k})}, \Xi^{(\mathbf{k})}\right)}{\left(\left[E^{(\mathbf{k})} + R_0^{(\mathbf{k})}\right]\Xi^{(\mathbf{k})}, \Xi^{(\mathbf{k})}\right)}. \tag{4.94}$$

When using Eqs.(4.88) and (4.89), we obtain the estimate

$$\sum_{\mathbf{k}\in K} \kappa^{(\mathbf{k})}\lambda_{\min}\left(B^{(\mathbf{k})}\right) \le \delta_d^2 \le \sum_{\mathbf{k}\in K} \kappa^{(\mathbf{k})}\lambda_{\max}\left(B^{(\mathbf{k})}\right) \tag{4.95}$$

where $\lambda_{\min}(.)$ and $\lambda_{\max}(.)$ are the minimum and maximum eigenvalues of matrix.

Let us consider now the second component $\delta_s^2$ error $\delta_f^2$, Eqs.(4.85) and (4.87). If we take into account Eq.(2.1) and perform transformations, we find that

$$\delta_s^2 = 2\left\{1 - \mathrm{Re}\sum_{\mathbf{k}\in K} \kappa^{(\mathbf{k})} \exp\left[i\Delta\alpha^{(\mathbf{k})}\right] \frac{\left(\mathrm{H}^{(\mathbf{k})}, \Xi^{(\mathbf{k})}\right)}{\left|\mathrm{H}^{(\mathbf{k})}\right\|\Xi^{(\mathbf{k})}\right|}\right\} \tag{4.96}$$

The choice of the phases $\Delta\alpha^{(\mathbf{k})}$ (left arbitrary up to now) can be done on the basis of the $\delta_s^2$ minimizing. The minimization uses an easily provable theorem.

The minimum value of the function

$$\mathrm{Re}\left(\sum_{\mathbf{k}\in K} z_{\mathbf{k}} \exp\left[i\Delta\alpha^{(\mathbf{k})}\right]\right)$$

with respect to the variables $\Delta\alpha^{(\mathbf{k})}$, $\mathbf{k}\in K$, is equal to

$$\sum_{\mathbf{k}\in K} \left|z_{\mathbf{k}}\right|.$$

and reached under the equality

$$\exp\left[i\Delta\alpha^{(\mathbf{k})}\right] = \frac{z_{\mathbf{k}}^*}{\left|z_{\mathbf{k}}\right|}$$

Thus, to minimize Eq.(4.96), one should choose $\tilde{\alpha}^{(\mathbf{k})}$ in the form of Eq.(4.35), where

$$\exp\left[i\Delta\alpha^{(\mathbf{k})}\right]=\frac{\left(\mathrm{H}^{(\mathbf{k})},\Xi^{(\mathbf{k})}\right)^*}{\left(\mathrm{H}^{(\mathbf{k})},\Xi^{(\mathbf{k})}\right)} \tag{4.97}$$

and the minimum value of $\delta_s^2$ equals

$$\delta_s^2=2(1-\cos\beta)=\sum_{\mathbf{k}\in K}\kappa^{(\mathbf{k})}2(1-\cos\beta^{(\mathbf{k})}), \tag{4.98}$$

where

$$\cos\beta=\sum_{\mathbf{k}\in K}\kappa^{(\mathbf{k})}\cos\beta^{(\mathbf{k})} \tag{4.99}$$

$$\cos\beta^{(\mathbf{k})}=\frac{\left|\left(\mathrm{H}^{(\mathbf{k})},\Xi^{(\mathbf{k})}\right)\right|}{\left|\mathrm{H}^{(\mathbf{k})}\right|\left|\Xi^{(\mathbf{k})}\right|}. \tag{4.100}$$

Note that Eq.(4.97) taken together with the above mentioned Eqs.(4.66), (4.69), and (4.31), allows the predistorted numbers $\tilde{\mu}^{(\mathbf{k})}$, Eq.(4.35), to be completely determined:

$$\tilde{\mu}^{(\mathbf{k})}=\sqrt{q}\;\mu^{(\mathbf{k})}\frac{\left(\mathrm{H}^{(\mathbf{k})},\Xi^{(\mathbf{k})}\right)^*}{\left|\left(\mathrm{H}^{(\mathbf{k})},\Xi^{(\mathbf{k})}\right)\right|}\cdot\frac{\left|\tilde{\Xi}^{(\mathbf{k})}\right|}{\left|\mathrm{H}^{(\mathbf{k})}\right|}$$

Equation specifying the relative error can as well be written in another form:

$$\delta_s^2=\left[2\sin\frac{\beta}{2}\right]^2=\sum_{\mathbf{k}\in K}\kappa^{(\mathbf{k})}\left[2\sin\frac{\beta^{(\mathbf{k})}}{2}\right]^2$$

The value $\beta^{(\mathbf{k})}$ denotes the angle between the vector of the required and the real mode compositions of the **k**-th branch measured in the $\left|L^{(\mathbf{k})}\right|$-dimensional Cartesian coordinates; $\beta$ is the branch-averaged angle.

Note that by virtue of Eq.(2.1) and the Cauchy-Bunyakovsky inequality

$$\left|\left(\mathrm{H}^{(\mathbf{k})},\Xi^{(\mathbf{k})}\right)\right|\le\left|\mathrm{H}^{(\mathbf{k})}\right|\left|\Xi^{(\mathbf{k})}\right|$$

there takes place the following inequality

$$0\le\left|\cos\beta^{(\mathbf{k})}\right|\le 1,\quad 0\le\left|\cos\beta\right|\le 1.$$

This means that the angles $\beta^{(\mathbf{k})}$, $\mathbf{k}\in K$, and $\beta$ are real and fall into the interval $\left[-\pi/2,\ \pi/2\right]$. Since in the Cauchy-Bunyakovsky inequality the exact equality takes place only for the collinear vectors $\mathrm{H}^{(\mathbf{k})}$ and $\Xi^{(\mathbf{k})}$, the parameter $\delta_s^2$ specifies the degree of non-collinearity between the vectors of really formed and required mode compositions. Using the angle $\beta$ itself $(|\beta|<\pi/2)$ we can also characterize an accuracy of the modes formation by a modan. The values β=0 and $\delta_s^2=0$ correspond to the formation of the required mode composition. On the contrary, the values $|\beta|=\pi/2$ and $\delta_s^2=1$ show the absolute absence of the required modes in the formed beam. Really operating modans can be produced by choosing appropriate sampling and the state-of-the-art technology for which the perturbations are small, that is, $\left\|H^{(\mathbf{k})}\right\|<<1$. If $\left\|H^{(\mathbf{k})}\right\|\to 0$, Eqs.(4.33) and (4.34) yield that $\left|\mathrm{H}^{(\mathbf{k})}\right|\to\left|\Xi^{(\mathbf{k})}\right|$ and $\left|\mathrm{H}^{(\mathbf{k})}-\Xi^{(\mathbf{k})}\right|\to 0$. Therefore, the first approximation in terms of a small parameter $\left\|H^{(\mathbf{k})}\right\|$ gives us

$$\beta^{(\mathbf{k})}=\frac{\left|\mathrm{H}^{(\mathbf{k})}-\Xi^{(\mathbf{k})}\right|}{\left|\Xi^{(\mathbf{k})}\right|}$$

$$\delta_s^2=\beta^2=\sum_{\mathbf{k}\in K}\kappa^{(\mathbf{k})}\left[\beta^{(\mathbf{k})}\right]^2 .$$

From Eq.(4.33) we find that

$$\left[\beta^{(\mathbf{k})}\right]^2=\frac{\left|H_r^{(\mathbf{k})}\Xi^{(\mathbf{k})}\right|^2}{\left|\Xi^{(\mathbf{k})}\right|^2}=\frac{\left(\Lambda^{(\mathbf{k})}\Xi^{(\mathbf{k})},\Xi^{(\mathbf{k})}\right)}{\left|\Xi^{(\mathbf{k})}\right|^2}, \tag{4.101}$$

where

$$\Lambda^{(\mathbf{k})}=H_r^{(\mathbf{k})*}H_r^{(\mathbf{k})} \tag{4.102}$$

As seen from Eqs.(4.101) and (4.102), the $\delta_s^2$ error depends on the specific values of the vector $\Xi^{(\mathbf{k})}$ of the required modal composition. Let us evaluate the range of variations of $\delta_s^2$ for the varying vectors $\Xi^{(\mathbf{k})}$ by estimating a non-negatively defined quadratic form of Eq.(4.101). Introducing eigenvalues of the matrix $\Lambda^{(\mathbf{k})}$, we obtain

$$\sum_{\mathbf{k}\in K}\kappa^{(\mathbf{k})}\left[\lambda_{\min}\left(H^{(\mathbf{k})}H^{(\mathbf{k})*}\right)\right]^{r+1}\le\delta_s^2\le\sum_{\mathbf{k}\in K}\kappa^{(\mathbf{k})}\left[\lambda_{\max}\left(H^{(\mathbf{k})}H^{(\mathbf{k})*}\right)\right]^{r+1} \tag{4.103}$$

If our task is only to obtain the required power distribution between modes without any regard to intermodal phase shifts, it appears expedient to modify the accuracy characteristic of the transverse-mode composition and choose another parameter $\delta_+^2$ instead of $\delta_s^2$, Eq.(4.87),

$$\delta_+^2 = \sum_{\mathbf{k}\in K} \kappa^{(\mathbf{k})} \left| \exp\left[i\Delta\alpha^{(\mathbf{k})}\right] \frac{\mathbf{H}_+^{(\mathbf{k})}}{\left|\mathbf{H}^{(\mathbf{k})}\right|} - \frac{\Xi_+^{(\mathbf{k})}}{\left|\Xi^{(\mathbf{k})}\right|} \right|^2$$

that characterizes the difference in power distributions between the modes without regard to phases. Thus, the vectors $\mathbf{H}_+^{(\mathbf{k})}$ and $\Xi_+^{(\mathbf{k})}$ are derived from $\mathbf{H}^{(\mathbf{k})}$ and $\Xi^{(\mathbf{k})}$, respectively, by means of taking modulus of each component. In this case optimal values of the phase shifts $\Delta\alpha_+^{(\mathbf{k})}$ and the $\delta_+$ characteristic are specified by Eqs.(4.97) through (4.100), with the vectors $\mathbf{H}^{(\mathbf{k})}$ and $\Xi^{(\mathbf{k})}$ replaced by $\mathbf{H}_+^{(\mathbf{k})}$ and $\Xi_+^{(\mathbf{k})}$ respectively.

For a singlemodal modan, Eq.(4.97) takes the form

$$\exp\left[i\Delta\alpha^{(\mathbf{p})}\right] = \frac{1+H_r^{(\mathbf{p})*}}{\left|1+H_r^{(\mathbf{p})}\right|} \tag{4.104}$$

The values $H^{(\mathbf{p})}$ and $H_r^{(\mathbf{p})}$ are specified in Eqs.(4.15) and (4.21). In that case the error of modal compositions is equal to zero:

$$\delta_s^2 = 0, \quad \Delta\beta = 0,$$

since a single mode has always its own composition. The total error $\delta_f^2$ is equivalent here to $\delta_d^2$ and, in view of Eqs.(4.95), (4.93), (4.89), can be estimated as follows:

$$\delta_f^2 = \sum_{\mathbf{p}\in K} \kappa^{(\mathbf{p})} \frac{Q^{(\mathbf{p})} - \left|H^{(\mathbf{p})}\right|^2}{\left|1+H^{(\mathbf{p})}\right|^2} \tag{4.105}$$

Computer-aided synthesis of modans requires: (1) to minimize the number of pixels directly affecting the time of synthesis and (2) avoid too precise resolution of a mask, in order to match to available facilities. On the other hand, modans corresponding to mode configurations of higher orders are better generated with a larger number of pixels and higher resolution. Estimates (4.85), (4.95), (4.103) and (4.105) show the way to satisfy the foregoing contradictory requirements. This can be made by studying the relations of the modan accuracy characteristics to the resolution $\delta_1, \delta_2$ and the parameters of the analyzed light beam and making choice of appropriate parameters capable of providing the preset accuracy. With reference to the example given in the end of section 4.4, the accuracy characteristic for the case of Gauss-Hermite modes is estimated by equation

$$\delta_f = \frac{\left(Q - H^2\right)^{1/2}}{1+H}, \tag{4.106}$$

where the numbers $Q$ and $H$ are specified by Eqs.(4.79) and (4.80). If $\delta \to 0$, we have that $N_\nu \to \infty$, $N_\delta \to \infty$, and (as should be expected) $\delta_f \to 0$. With the increasing spatial carrier frequency ν (decreasing $N_\nu$) the energy efficiency falls, and the mean-square deviation $\delta_f$ rises (see table 4.4); this means simultaneous deterioration of the quality of the formed complex amplitude distribution with the decreasing portion of incident light flux. Thus, in the presence of sampling, one should minimize the spatial frequency ν. However, one should bear in mind that the lower bound of ν is determined by the conditions of separation of zero and first diffraction orders.

As seen from analysis of Eqs.(4.82) and (4.106) and Fig. 4.10, the energy efficiency falls with increasing order ($p+l$) of the $\psi_{pl}$ mode, whereas the error in mode formation rises. One should also take into account that the rise of the order ($p+l$) of a mode characterized by the width of $\sigma\sqrt{p+0.5} \times \sigma\sqrt{l+0.5}$ is limited by the size $2a \times 2b$ of a modan.

The task of analysis of a mode composition leads to different parameters of accuracy. They are related to the coefficients of the expansion of the light field analyzed.

In the presence of perturbations, the beam (2.45) - (2.46) reconstructed from a modan, see Eqs.(2.47), incorporates the perturbed beam $\hat{W}_{\mathbf{n}}^{(\mathbf{k})}$, Eq.(4.3) instead of the partial beams $W_{\mathbf{n}}^{(\mathbf{k})}$, and the complex amplitudes, Eq.(3.7), are replaced by

$$\hat{w}^{(\mathbf{k})} = \left(w, \hat{W}_{\mathbf{n}}^{(\mathbf{k})}\right).$$

This means that the mode representations already involve "perturbed" coefficients $\hat{w}_{\mathbf{p}}$ being only an approximation of the mode coefficients $w_{\mathbf{p}}$.

[Tikhonov, 1986] have studied the errors in representation of the functions by orthogonal expansions with ill-conditioned coefficients. When generating modans as computer-synthesized elements, there arises a specific type of the measured coefficients' perturbations, in which physics of the problem and the type of perturbed base functions produced by sampling and other perturbations of modans are reflected. Hereafter we formulate the strategy of investigation of how perturbations affect an accuracy of the mode composition analysis. The error in the **k**-th branch is found from the following equation

$$\Delta_a^{(\mathbf{k})} = \left|\frac{1}{\sqrt{q}}\hat{w}^{(\mathbf{k})} - w^{(\mathbf{k})}\right| = \left|\left(w, \frac{1}{\sqrt{q}}\hat{W}_{\mathbf{n}}^{(\mathbf{k})} - W_{\mathbf{n}}^{(\mathbf{k})}\right)\right| \tag{4.107}$$

that accounts of the decrease in the energy efficiency due to perturbations. Total error in the mode composition analysis throughout the set of the branches $\mathbf{k} \in K$ is specified as a sum of single-branch errors

**Table 4.4**
**Modan's accuracy $\delta_f$ versus a spatial carrier frequency ($p+l=10$)**

| $N_\sigma$ | $N_\nu$ : | 4 | 6 | 8 | 10 | 20 | 50 |
|---|---|---|---|---|---|---|---|
| 10 | | 0.509 | 0.343 | 0.270 | 0.23 | 0.164 | 0.141 |
| 100 | | 0.483 | 0.311 | 0.230 | 0.183 | 0.092 | 0.039 |

Modified from Golub M.A., Sisakyan I.N., Soifer V.A., Computer generated optical elements for fiber's mode selection and launching, *Proceedings SPIE, OE/Fibers'90*, 1365, 156, 1990.

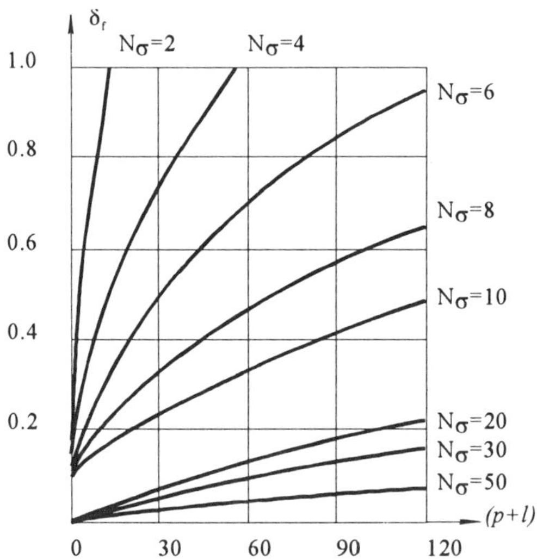


**Figure 4.10.** The degree of the beam changes caused by parasitic modes due to the Gauss-Hermite modan sampling. $\delta_f$- modan's accuracy; $(p+l)$ - order of the required mode.

$$\Delta_a^2 = \sum_{\mathbf{k}\in K} \left[\Delta_a^{(\mathbf{k})}\right]^2,$$

and the relative error is determined as

$$\delta_a^2 = \frac{\Delta_a^2}{\sum_{\mathbf{k}\in K} \left|w^{(\mathbf{k})}\right|^2} \tag{4.108}$$

In view of Eq.(4.83), value (4.107) admits the following estimate

$$\Delta_a^{(\mathbf{k})} \leq \|w\| \cdot \left\| \frac{1}{\sqrt{q}} \hat{W}_{\mathbf{n}}^{(\mathbf{k})} - W_{\mathbf{n}}^{(\mathbf{k})} \right\| = \|w\| \cdot \Delta_f^{(\mathbf{k})}$$

Eqs.(3.7) and (4.84), allow the relative error to be written as follows

$$\delta_a^2 \leq \frac{\delta_f^2}{\sum_{\mathbf{k} \in K} \kappa^{(\mathbf{k})} \left| \left( \frac{w}{\|w\|}, \frac{\xi^{(\mathbf{k})}}{\left|\Xi^{(\mathbf{k})}\right|} \right) \right|^2}, \qquad \kappa^{(\mathbf{k})} = \frac{1}{L}$$

The value $\delta_f^2$ was introduced earlier (see Eqs.(4.85), (4.87), (4.98)) as the characteristic of formation accuracy.

## 4.6 ACCURACY OF MODAL SPECTRAL TRANSFORMS

Estimation of the errors in mode transforms is based on essentially different characteristics of accuracy as compared with the mode formation and analysis. Since a finite number of the series coefficients are to specify the field uniquely, we should here estimate the truncation error in the multimodal beam representation in terms of modes. In other words we have to evaluate the error in the representation of a series through the finite number of its terms, under the condition that there occur perturbations of base functions or of the coefficients. In the stage of a direct mode transform, perturbations effect results in the calculation of the perturbed coefficients

$$\hat{w}^{(\mathbf{p})} = \sqrt{\frac{E_0 \varepsilon_{\mathbf{n}} q}{L}} \cdot \frac{1}{1 + H^{(\mathbf{p})}} \hat{w}_{\mathbf{p}}$$

$$\hat{w}_{\mathbf{p}} = \int_G w(\mathbf{x}) \varphi_{\mathbf{p}}(\mathbf{x}) d^2\mathbf{x}, \quad \mathbf{p} \in K \tag{4.109}$$

instead of the corresponding coefficients $w^{(\mathbf{p})}$ of series (3.16) and (3.5).

From the results of the direct transform we can restore the field

$$w(\mathbf{x}) = \sum_{\mathbf{p}} w_{\mathbf{p}} \psi_{\mathbf{p}}(\mathbf{x})$$

from a perturbed expansion with a finite number of terms

$$\zeta(\mathbf{x}) = \sum_{\mathbf{p} \in K} \frac{w_{\mathbf{p}}}{1 + H^{(\mathbf{p})}} \psi_{\mathbf{p}}(\mathbf{x}). \tag{4.110}$$

Note that series (4.110) are matched to approximation (4.11).

In order to bring the scale of the values of the perturbed coefficients $\hat{\mathrm{w}}^{(\mathbf{p})}$ to that of the $w^{(\mathbf{p})}$ coefficients, Eq.(3.16), equation (4.110) was normalized by the factor $1/\sqrt{q}$. Note that Eq.(4.110) includes the information about the $w(\mathbf{x})$ function in terms of both generalized-spectral components and the accuracy of obtaining each component. The error of the replacement

$$w(\mathbf{x}) \cong \zeta(\mathbf{x})$$

can be defined as follows

$$\Delta_m^2 = \|\zeta - w\|^2 = \sum_{\mathbf{p}\in K} \left| \frac{\hat{w}_{\mathbf{p}}}{1+H^{(\mathbf{p})}} - w_{\mathbf{p}} \right|^2 + \sum_{\mathbf{p}\notin K} \left| w_{\mathbf{p}} \right|^2 =$$

$$= \sum_{\mathbf{p}\in K} \left| \left( w, \frac{\varphi_{\mathbf{p}}}{1+H^{(\mathbf{p})*}} - \psi_{\mathbf{p}} \right) \right|^2 + \|w\|^2 - \sum_{\mathbf{p}\in K} \left| w_{\mathbf{p}} \right|^2 \le$$

$$\le \|w\|^2 \sum_{\mathbf{p}\in K} \left\| \frac{\varphi_{\mathbf{p}}}{1+H^{(\mathbf{p})*}} - \psi_{\mathbf{p}} \right\|^2 + \|w\|^2 - \sum_{\mathbf{p}\in K} \left| w_{\mathbf{p}} \right|^2$$

The relative error of the direct mode transform is

$$\delta_m^2 = \frac{\Delta_m^2}{\|w\|^2} = \delta_0^2 + \delta_\pi^2$$

where, in view of Eq.(4.18), we have

$$\delta_\pi^2 = \sum_{\mathbf{p}\in K} \frac{Q^{(\mathbf{p})} - \left| H^{(\mathbf{p})} \right|^2}{\left| 1+H^{(\mathbf{p})} \right|^2}, \tag{4.111}$$

and $\delta_0^2$ can be found from Eq.(3.13). The value $\delta_\pi^2$ describes the error resulted from the inaccuracy in the calculation of the coefficients and increases with the number $N_c$ of the spectral components $\mathbf{p} \in K$. The value $\delta_0^2$ characterizes the mean-square error in representing the function by a finite number of spectral components and decreases with the increasing number of the components. This means that the $\delta_m^2$ function has the minimum value with respect to the variable $N_c$.

Let us examine how perturbations affect the operation of a stage for the inverse mode transform. In the presence of perturbations, the modan yields, instead of Eq.(3.18), the following transfer function:

$$\hat{\Gamma}_{\mathbf{n}}(\mathbf{x}') = \frac{1}{w_0} \sqrt{\frac{E_0 \varepsilon_{\mathbf{n}} q}{L}} \sum_{\mathbf{p}'\in K} \frac{\varphi_{\mathbf{p}'}(\mathbf{x}')}{1+H^{(\mathbf{p}')*}} \exp\left[ i2\pi \left( \boldsymbol{\nu}^{(\mathbf{p}')} + \boldsymbol{\nu}_{\mathbf{n}} \right) \mathbf{x}' \right]$$

that is, it reconstructs the field

$$\sqrt{\frac{E_0\varepsilon_{\mathbf{n}}q}{L}}\left\{\hat{W}_L(\mathbf{x}')+\sum_{\mathbf{p}'\in K}W_{\mathbf{p}}\sum_{\mathbf{p}'\neq\mathbf{p}}\frac{\varphi_{\mathbf{p}'}(\mathbf{x}')}{1+H^{(\mathbf{p}')*}}\exp\left[i2\pi\left(\boldsymbol{\nu}^{(\mathbf{p}')}-\boldsymbol{\nu}^{(\mathbf{p})}\right)\mathbf{x}'\right]\times\right.$$
$$\left.\times\exp\left[i2\pi\left(\boldsymbol{\nu}^{(\mathbf{p})}+\boldsymbol{\nu}_{\mathbf{n}}\right)\mathbf{x}'\right]\right\} \qquad (4.112)$$

where

$$\hat{W}_L(\mathbf{x}')=\sum_{\mathbf{p}\in K}W_{\mathbf{p}}\frac{\varphi_{\mathbf{p}}(\mathbf{x})}{1+H^{(\mathbf{p})*}}$$

If the diffraction orders in Eq (3.15) are separated (Eq (2.4)) to such an extent that $\varphi_{\mathbf{p}'}(\mathbf{x}')$ do not overlap, the first sum $\hat{W}_L$ in Eq.(4.112) corresponds to the field that has been really obtained. The mean-square deviation of the inverse mode transform is equal to;

$$\delta_t^2=\frac{1}{E}\left\|\hat{W}_L-W\right\|^2.$$

Using the representation

$$\hat{W}_L-W=\sum_{\mathbf{p}\in K}W_{\mathbf{p}}\left[\frac{\varphi_{\mathbf{p}}}{1+H^{(\mathbf{p})*}}-\psi_{\mathbf{p}}\right]-\sum_{\mathbf{p}\notin K}W_{\mathbf{p}}\psi_{\mathbf{p}},$$

the mode orthogonality in Eq.(1.3), and Eq.(4.18), we find the estimate

$$\delta_t=\delta_\gamma(L)+\delta_h(L), \qquad (4.113)$$

where

$$\delta_\gamma^2(L)=1-\sum_{\mathbf{p}\in K}\frac{\left|W_{\mathbf{p}}\right|^2}{E_\gamma} \qquad (4.114)$$

$$\delta_h^2(L)=\frac{1}{E}\left\|\sum_{\mathbf{p}\in K}W_{\mathbf{p}}\left[\frac{\varphi_{\mathbf{p}}(\mathbf{x})}{1+H^{(\mathbf{p})*}}-\psi_{\mathbf{p}}\right]\right\|^2=\frac{1}{E_\gamma}\sum_{\mathbf{p},\mathbf{p}'\in K}M_{\mathbf{pp}'}W_{\mathbf{p}}W_{\mathbf{p}'}^* \qquad (4.115)$$

where

$$M_{\mathbf{pp}'}=\frac{Q_{\mathbf{pp}'}^{(\mathbf{p},\mathbf{p}')*}+\left|H^{(\mathbf{p})}\right|^2\delta_{\mathbf{pp}'}-H^{(\mathbf{p})*}H_{\mathbf{pp}'}^{(\mathbf{p}')}-H^{(\mathbf{p}')}H_{\mathbf{pp}'}^{(\mathbf{p})*}}{\left[1+H^{(\mathbf{p})*}\right]\left[1+H^{(\mathbf{p}')}\right]} \qquad (4.116)$$

The matrix elements $Q$ and $H$ have been specified in section 4.1. The error $\delta_h(L)$ allows the estimate to be done through the eigenvalues of a self-conjugate and positively determined matrix

$$M_L=\left[M_{\mathbf{pp}'},\ \mathbf{p},\mathbf{p}'\in K\ \right]$$

of quadratic form, Eqs.(4.115) and (4.116).
The estimate can be written down in the form

$$\lambda_{\min}(M_L)\sum_{\mathbf{p}\in K}\frac{|W_{\mathbf{p}}|^2}{E_\gamma}\le\delta_n^2(L)\le\lambda_{\max}(M_L)\sum_{\mathbf{p}\in K}\frac{|W_{\mathbf{p}}|^2}{E_\gamma}, \tag{4.117}$$

$\lambda_{\min}(\cdot)$ and $\lambda_{\max}(\cdot)$ are the minimal and maximum eigenvalues of the $M_L$ matrix.

From the first equality in Eq.(4.115) and Eq.(4.18) we find that the more rough estimate of $\delta_h^2$ is as follows

$$\delta_n^2(L)\le\sum_{\mathbf{p}\in K}\frac{Q^{(\mathbf{p})}-\left|H^{(\mathbf{p})}\right|^2}{\left|1+H^{(\mathbf{p})}\right|^2} \tag{4.118}$$

From the first equality in (4.115) and from Eq.(4.18) we find that the more rough but simpler estimate of $\delta_h^2$ is as follows

$$\delta_h^2(L)\le\sum_{\mathbf{p}\in K}\frac{Q^{(\mathbf{p})}-\left|H^{(\mathbf{p})}\right|^2}{\left|1+H^{(\mathbf{p})}\right|^2}.$$

# Chapter 5

# MODAN'S DEVELOPMENT AND EXPERIMENTAL INVESTIGATION

## 5.1 COMPUTER GENERATION OF MODANS BY DIFFRACTIVE OPTICS TECHNOLOGY

This chapter describes more experimental engineering issues rather than theoretical ones. Here we discuss the problems of modans' implementation by computer-aided diffractive optics technology. Results of various experiments on modes reconstruction from modans are presented.

Looking from the point of view of digital holography we should record spatial filters matched to the mode $\psi_p(\mathbf{x})$ or to the superposition of several modes if required. Due to the complex (amplitude-phase) character of modes, appropriate spatial filters must, in a general case, have the complex transfer function. In such cases as Gauss-Laguerre or Gauss-Hermite modes a real-value transfer function with alternating sign occurs.

**A. Amplitude modans coded by additive constant**

The results development and investigation of amplitude spatial filters with zero carrier frequency synthesized with the aid of computer and matched to real modes of coherent radiation have been reported in the papers [Golub, 1982], [Golub, 1984]. Addition of a constant as in Eq.(2.43) have been used for coding real-value transfer function into positive one. Figures 5.1 through 5.3 illustrate the masks for the synthesized modans. In the course of computer-aided generation we have used 256x256 pixel on each mask with the amplitude transmittance ranging from 0.1 to 0.8 and with 256 graduations of an optical density.

Figure 5.2 depicts an amplitude multimodal singlebeam modan that enables us to form an standard multimodal beam comprising Gauss-Laguerre modes of Eq.(1.108). The mode composition of the beam is specified in Table 5.1.

The amplitude modans described above make it possible to solve the problems of a mode selection. They were the first ones generated but suffer from a number of shortcomings. Such modans characterized by a constant shift do not allow for optical separation of the useful and auxiliary diffraction orders, because they do not contain a spatial carrier frequency. Thus they do not allow us to observe a correlation peak in an explicit form in the course of optical analysis of modal power distribution. In addition, some digital processing of measured data is needed in this case of additive- constant amplitude modan.

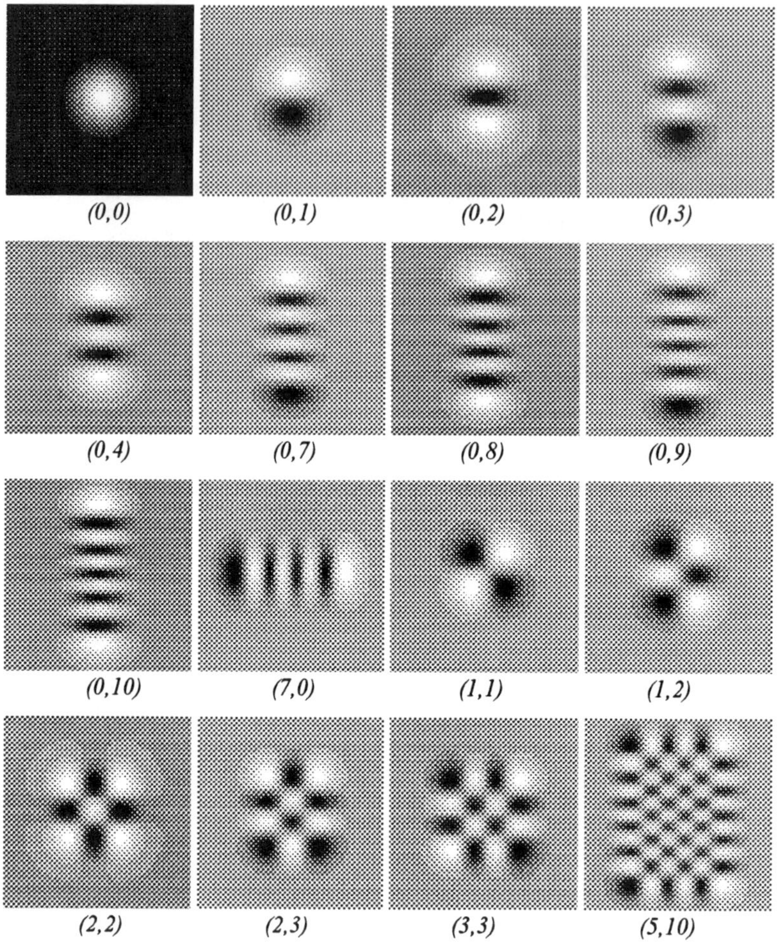


**Figure 5.1.** Magnified pictures of the amplitude spatial filters matched to Gauss-Hermite modes (encoding is made by adding a constant). (Modified from Golub M.A., Sisakyan I.N., Soifer V.A., Mode selection of laser radiation by computer-generated optical elements, *Optics and lasers in Engineering*, 15(5), 314, 1991).

**B. Amplitude modans with non-zero carrier**

Amplitude modans characterized by non-zero cosinusoidal spatial carrier frequency have been proposed by Karpeev S.V., Uvarov G.V. and reviewed in [Sisakyan, 1985].

Modans described by [Bartelt, 1983] have been implemented by means of the A.W.Lohmann's binary cells (see [Brown, 1966]). The modans have been implemented on the basis of Eq.(2.43) specifying an amplitude coding with the carrier. A manufacture of both single-beam (see [Golub, 1988]) and two-beam (see [Adzhalov, 1990]) modans has been reported, the latter having frequencies as (with $\nu$=10 lines/mm) and shown in Fig. 5.4.

**Table 5.1**
**Modal composition of the beam**

| $(p,l)$ | (0,0) | (1,1) | (0,5) | (7,0) | (5,10) |
|---|---|---|---|---|---|
| $\xi_{pl}/\xi_{00}$ | 1.0 | 0.8 | 0.7 | 0.6 | 0.1 |

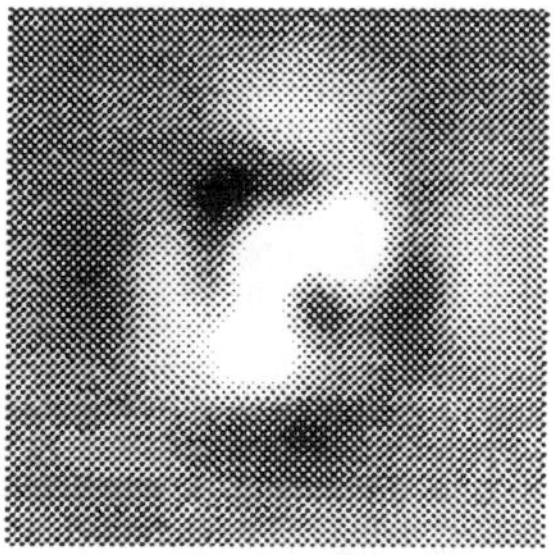

**Figure 5.2.** Amplitude spatial filter designed to form a reference multimodal Gauss-Hermite beam with the coefficients of Table 5.1.

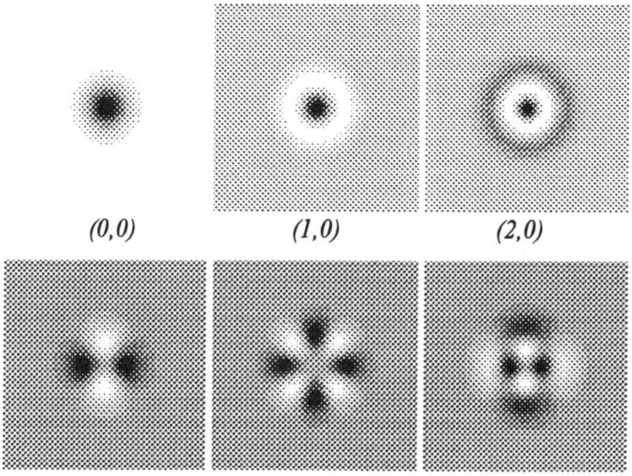


**Figure 5.3.** Magnified pictures of the amplitude spatial filters matched to Gauss-Laguerre modes. (Modified from Golub M.A., Sisakyan I.N., Soifer V.A., Mode selection of laser radiation by computer-generated optical elements, *Optics and lasers in Engineering*, 15(5), 314, 1991).

These modans are able to meet all the requirements on a mode selection, but possess a low energy efficiency coming up to a fraction of a per cent only.

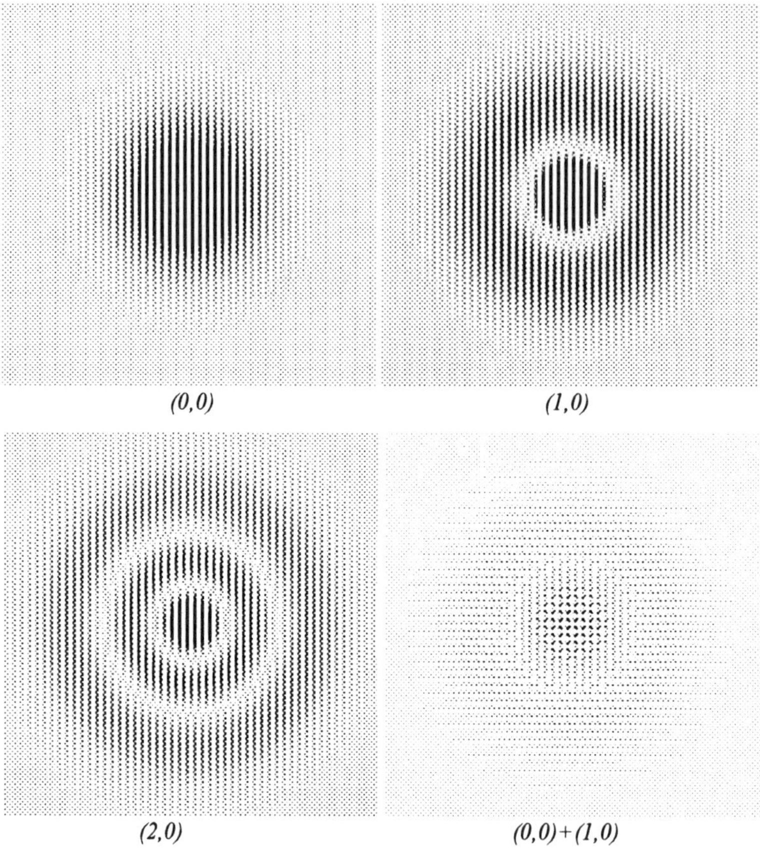


**Figure 5.4.** Magnified masks of single-beam and two-beam amplitude modans. The "+" sign denotes adding the modal functions. (Modified from Adzhalov V.I., Golub M.A., Karpeev S.V., Sisakyan I.N., Soifer V.A., Multi-channel computer optics components matched to mode groups, *Soviet Journal of Quantum Electronics*, 20(2), 136, 1990).

**C. Modans with amplitude-and-phase transfer**

One of good ways to implement amplitude-and-phase modans is using a "ROACH-technology" [Chu, 1974] for manufacturing a "sandwich." A more simple variant of a "binary sandwich" has been reported by [Berdague, 1982]. To ensure the required binary amplitude transmittance, complicated diaphragms were cut along the boundaries of the corresponding zeros of mode functions. Special gas-filled cells in which the controlled pressure determined the phase shifts and ensured the binary phase transfer of 0 or $\pi$, were employed. However, such filters are difficult to realize and do not allow of reaching the high degree of selectivity in exciting the modes, which is due to the binary character of the transfer function.

Any types of sandwich filters have limits in terms of their energy efficiency because of an amplitude transmission involved.

[Bartelt, 1985] have described so called "tandem" phase optical elements implemented as two separate and spaced optical elements. They are able to ensure the optimum energy efficiency. The first phase optical element forms an amplitude distribution. The second one performs the phase correction. However, one should bear in mind some difficulties of mutual alignment of tandem optical elements.

**D. Phase modans**

High energy efficiency considerations necessitate the realization of modans characterized by the phase - only transfer. Such modans are most promising ones and nearest to the things called "optical elements". Transformation of a complex transfer into a phase-only one is dealt with in a digital holography and called the transfer function "encoding". The diffractive control of amplitude in a beam cross-section that is realized by directing a portion of the power into inoperable diffraction orders, suits to the encoding processes best of all in the case of modans. Appropriate methods such as a method of synthetic coefficients [Chu, 1974] or the Kirk-Jones method (see [Kirk, 1971]), yield modans in the form of the phase diffraction gratings characterized by the curvilinear configuration and complex profile of grooves.

Computer-aided synthesis of phase modans involves the following stages: (1) the calculation of the matrix of the pixels of transfer function; (2) encoding of the complex pixels or real pixels of an alternating sign into the phase-only pixels; (3) formation of the arrays of digital codes for masks; (4) generation of masks; (5) etching or formation of the phase microrelief using the masks. Initial masks were generated on a drum photowriter with the resolution of 10 μm. In this case, 2048 x 2048 standard pixels of the transfer function were employed.

Two types of technology were used to realize a microrelief: the technology of the continuous microrelief and the method of a staircase microrelief.

In the framework of continuous microrelief approach, initial half-tone masks having 256 gradations of optical density ranging from 0.2 to 2.0 D units were generated. One-step manufacture of a microrelief was performed using a bleaching technology as well as continuous-level photo-resist approach. Bleaching is the simplest technology with high resolution of 200 lines/mm, but makes the microrelief surface too rough. Continuous-level photoresists ensure a good resolution but produce small height of a microrelief (not more than 0.1-0.7 μm). Photopolymeric materials that have been studied yield optically smooth surface of a microrelief with a maximum height of 10 μm. However, they have limitations on a spatial frequency up to 10-20 lines/mm and are not applicable for modans now.

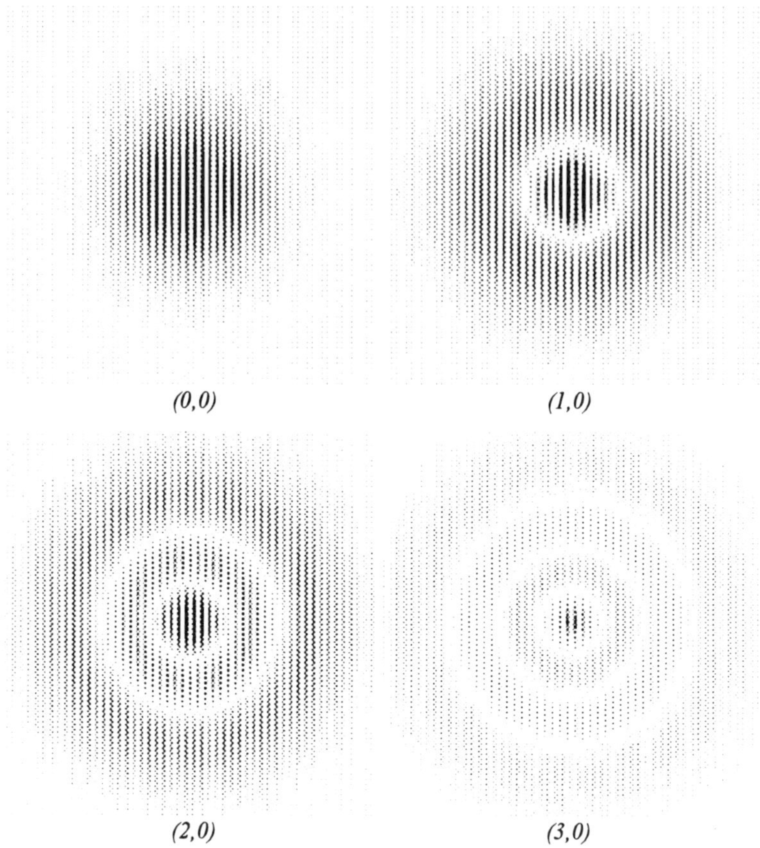


**Figure 5.5.** Magnified mask for a singlemodal phase modans of Gauss-Laguerre with the carrier of 10 lines/mm.

Using bleaching technology, we produced single-beam singlemodal Gauss- Laguerre modans characterized by the parameter $\sigma$= 0.65 mm and by the carrier frequency $\nu$ = 10 lines/mm (see Fig. 5.5).

Much more advanced technology was used for generation a six-beam modan reconstructing modes with the indices (0,0), (1,0), (2,0), (3,0), (0,1), and (0,2), with the carrier frequency of mode separation of 20 lines/mm and with the carrier frequency of 40 lines/mm for encoding an amplitude into a phase. The size of modan was 5.1 mm (see Fig. 5.6). A staircase microrelief was generated using multi-stage photolithography technology with the projection exposure ($10^{\times}$) and precise alignment. A set of four masks was manufactured for each modan. A polished glass plate of the 2 mm thickness was used as a substrate and has been put through four consecutive stages of the wet etching by hydrofluoric acid. Binary-phase modans have been also realized by S.V. Karpeev [Bereznyi, 1991].

**Figure 5.6.** Magnified masks for a Gauss- Laguerre six-beam singlemodal modan. The carrier frequencies are 20 lines/mm for reconstructing angular-separated modes (0,0) (1,0) (2,0) (3,0) (0,1) (0,2) and 40 lines/mm for complex-to phase coding. (Modified from Golub M.A., Sisakyan I.N., Soifer V.A., Mode selection of laser radiation by computer-generated optical elements, *Optics and lasers in Engineering*, 15(5), 314, 1991).

Its generation was based on detour phase and relative groove width modulation as described in section 2.5. Binary phase Gauss-Laguerre modan matched to the mode $(p,l) = (2,0)$ and with the carrier $\nu$ of 20 lines/mm was in generation. A photowriter with a resolution of 25 μm was used to generate a binary mask (see Fig. 5.7). Then a contact copy was made from the photofilm mask onto a chromium oxide coating on a glass substrate. A photoresist was exposed to ultra- violet light with $10^{\times}$ magnification. A special objective characterized by a numerical aperture of NA = 0.35 and by the resolution of about 1000 $mm^{-1}$ was employed here.

A Linnik microinterferometer was used to check the obtained phase microrelief. Simple one-stage manufacturing technology as well as the possibility of easily organizing a large-scale production are the benefits of the binary modans.

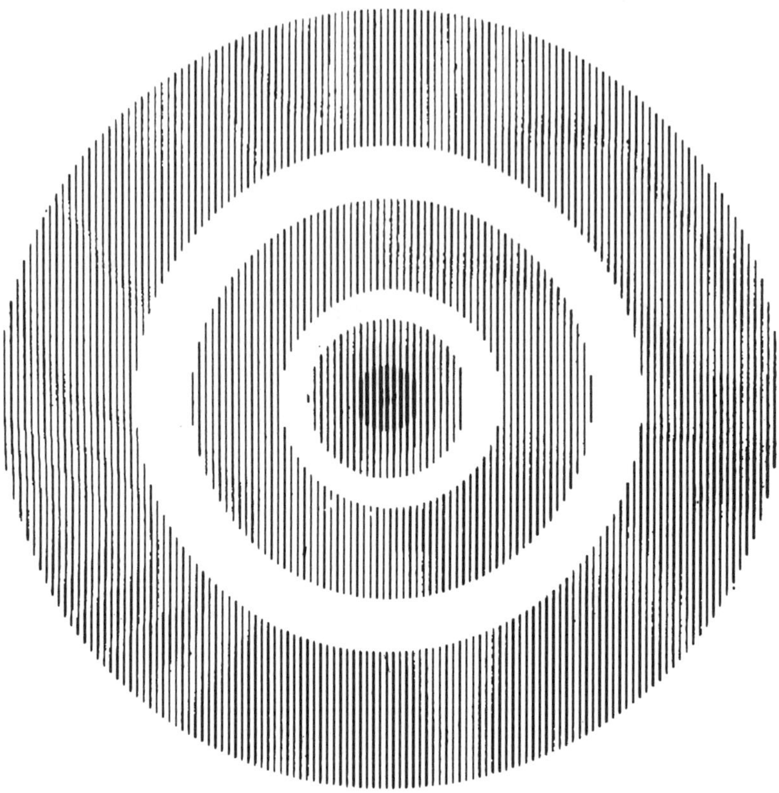

**Figure 5.7.** Binary mask of a holographic modan matched to (2,0) Gauss- Laguerre mode; carrier spatial frequency is 20 lines/mm. (Modified from Bereznyi A.E., Karpeev S.V., Uspleniev G.V., Computer generated holographic optical elements produced by photolithography, *Optics and Lasers in Engineering*, 15(5), 331, 1991).

In comparison to the binary modans without the carrier (see section 3.13) described by [Berdague, 1982], we have here the formation of the desired mode in a single diffraction order only.

Experimental investigation of modans included the following stages:

(1) Investigation of modan's performances, including the energy efficiency and the intensity distribution in the reconstructed modes (see section 5.2);

(2) Qualitative investigation of the operating ability of two-beam modans (see section 5.3);

(3) Measurements of power distribution between modes in a graded-index fiber waveguide excited by the Gaussian beam (see section 5.4);

(4) Selective excitation of modes of a graded-index fiber (see sections 5.5 and 5.6).

## 5.2 MEASUREMENTS OF EFFICIENCY AND OPERATING PERFORMANCES OF SINGLE-BEAM MODANS

**A. Measurements of the energy efficiency of modans**

Optical set-up schematically shown in Fig. 5.8 was used. A modan *M* limited by a diaphragm *D*, diameter 5 mm, was illuminated by the plane wave going from a laser *L* through the beam-expander *P*. The light flux directed by the modan into the first diffraction order was measured by a photoreceiver with the aperture $D_1$. A flux of incident light beam was measured by another photoreceiver *FS* placed right behind to the modan *M*.

Continuous-level amplitude and phase modans with the spatial carrier were studied. The results presented in Table 5.2 show that phase modans give the improvement in the energy efficiency by 8-9-fold, as compared with the amplitude modans; this value being close to the theoretical one of 11.1 derived from the equations in section 4.4.

**B. Measurements of the quality of mode image reconstruction**

To check experimentally the quality of the formation of separate Gauss-Laguerre modes, we used the set-up depicted in Fig. 5.9. Plane light beam from the beam-expander *P* passes the diaphragm *D* and comes to the modan *M,* where it is transformed by the Fourier stage, using the objective *0*. In the first diffraction order of a focal plane, the field (after a diagram *G* with the diameter 1 mm) comes to a photoreceiver *FS* implemented as an array of CCDs (Charge-Coupled Devices), each cell having the 15 x 15 μm size. The replacement of optical elements is performed mechanically with the help of special cassette.

Intensity distributions measured in a focal plane in the case of phase modans with the carrier (see Fig. 5.5) normalized on the total power are presented in Fig. 5.10. Dashed lines in Fig. 5.10 depict theoretical intensity distributions for the orthonormalized Gauss-Laguerre modes. From the analysis of Fig. 5.10 we can see that the experimental curves are in a good agreement with the theoretical ones. Figure 5.11 shows the diffraction pattern produced by the modan that has been described in the paper [Bereznyi, 1991] and depicted in Fig. 5.7. The power in the corresponding diffraction orders is illustrated in Fig. 5.11. For the sake of comparison, note that corresponding values typical of the filters characterized by the continuous phase microrelief and produced on the bleached gelatine layers (see [Golub, 1988]), amount to 0.32 and 0.02 for the zero and the first diffraction orders, respectively. Thus, the efficiency of the synthesized filters is seen to be five times as high as that of the filters that were utilized earlier.

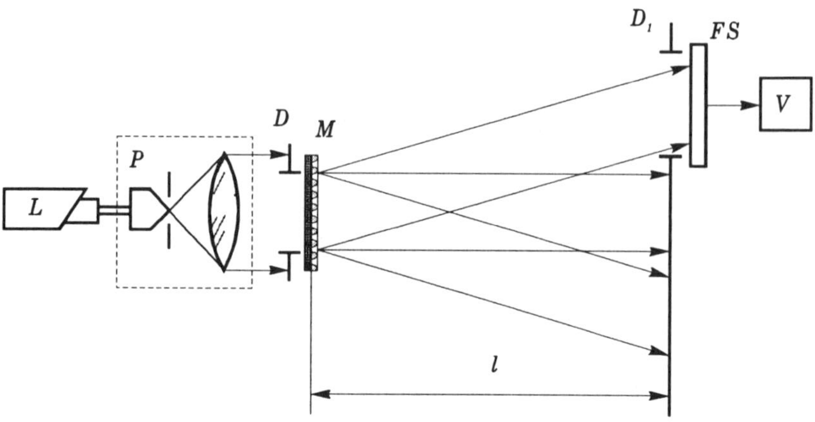


**Figure 5.8.** Optical set-up for measuring the energy efficiency of modans. $L$ is *He-Ne* laser, $P$-beam expander, $D$-diaphragm, $M$-modan, *FS*-photoreceiver with aperture $D_1$, V-electric signal indicator. (Modified from Golub M.A., Sisakyan I.N., Soifer V.A., Uvarov G.V., Optical components for analysis and formation of the transverse mode composition, *Soviet Journal of Quantum Electronics*, 19(4), 543, 1989).

**Table 5.2**
**Relative characteristics of phase and amplitude modans. ($p$,$l$) is the number of mode, $\varepsilon_{10}^{(pl)}$-energy efficiency for the first diffraction order, $|w_{pl}|^2$-modal power**

| $p$ | $l$ | $\varepsilon_{10}^{(pl)} / \varepsilon_{10}^{(pl)}{}_{(amplitude)}$ | | $\lvert w_{pl}\rvert^2 / \lvert w_{pl(amplitude)}\rvert^2$ | |
|---|---|---|---|---|---|
| | | **theory** | **experiment** | **theory** | **experiment** |
| 0 | 0 | 11.1 | 9.3 | 11.1 | 7.2 |
| 1 | 0 | 11.1 | 8.1 | 11.1 | 7.9 |
| 2 | 0 | 11.1 | 8.7 | 11.1 | 7.3 |

Modified from Golub M.A., Sisakyan I.N., Soifer V.A., Uvarov G.V., Optical components for analysis and formation of the transverse mode composition, *Soviet Journal of Quantum Electronics*, 19(4), 543, 1989.

Figure 5.12 illustrates the theoretical (dashed line) and experimental (solid line) plots for the intensity distributions, which occur in the first (useful) diffraction order and are obtained with a scanning photodiode, with the diaphragm of 100 μm diameter. Theoretical and experimental results are seen to be in a good agreement.

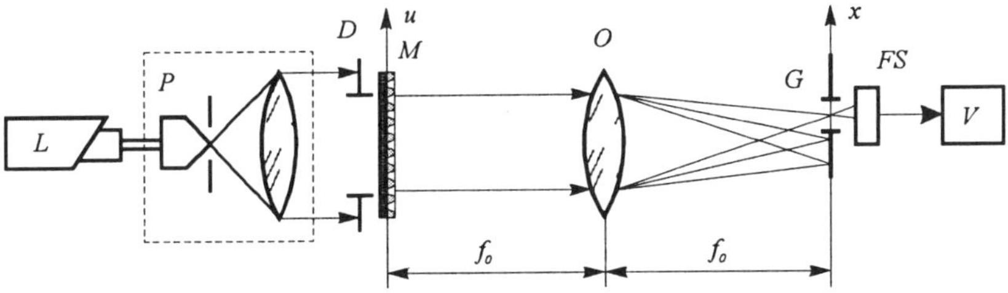


**Figure 5.9.** Optical set-up for modes reconstruction from holographic modans. *L* is He-Ne laser, *D*-diaphragm, *M*-modan, *P*-beam expander, *O*-objective with focal length $f_0$, *PS*-photoreceiver with aperture *G*, *V*-electric signal indicator. (Modified from Golub M.A., Sisakyan I.N., Soifer V.A., Uvarov G.V., Optical components for analysis and formation of the transverse mode composition, *Soviet Journal of Quantum Electronics*, 19(4), 543, 1989).

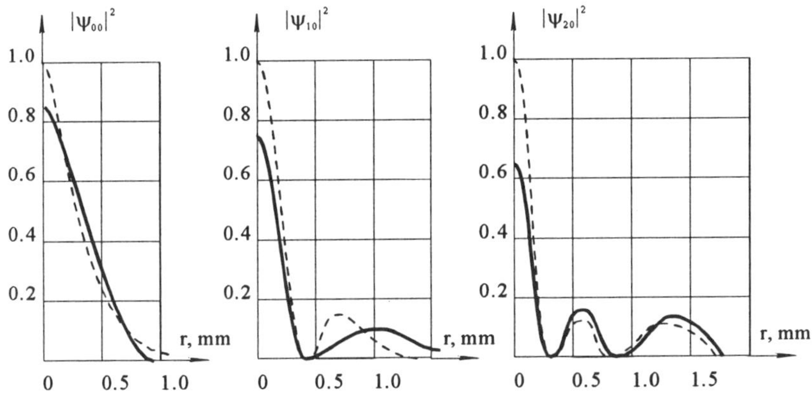


**Figure 5.10.** Normalized intensity distribution for the Gauss-Laguerre modes $\psi_{pl}$, $\sigma$=0.65 mm, $l$=0, $p$=0,1,2 (solid line - experimental distribution, and dashed line - a theoretical one). (Modified from Golub M.A., Sisakyan I.N., Soifer V.A., Uvarov G.V., Optical components for analysis and formation of the transverse mode composition, *Soviet Journal of Quantum Electronics*, 19(4), 543, 1989).

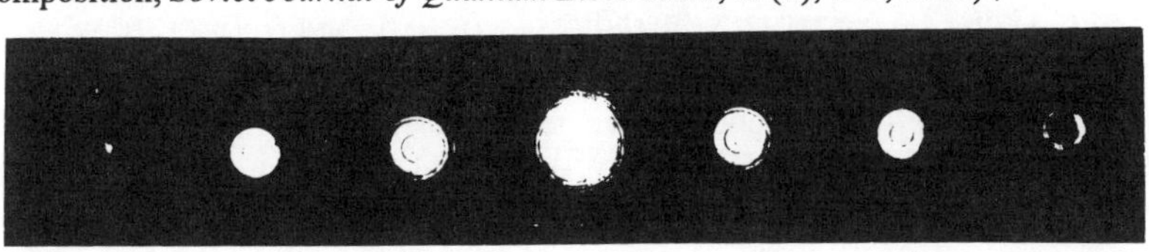

| | | | | | | | |
|---|---|---|---|---|---|---|---|
| *Order:* | -3 | -3 | -1 | 0 | +1 | +2 | +3 |
| *Power:* | 0.006 | 0.03 | 0.1 | 0.65 | 0.1 | 0.03 | 0.006 |

**Figure 5.11.** Photograph of the diffraction pattern produced by a binary-phase modan. (Modified from Bereznyi A.E., Karpeev S.V., Uspleniev G.V., Computer generated holographic optical elements produced by photolithography, *Optics and Lasers in Engineering*, 15(5), 331, 1991).

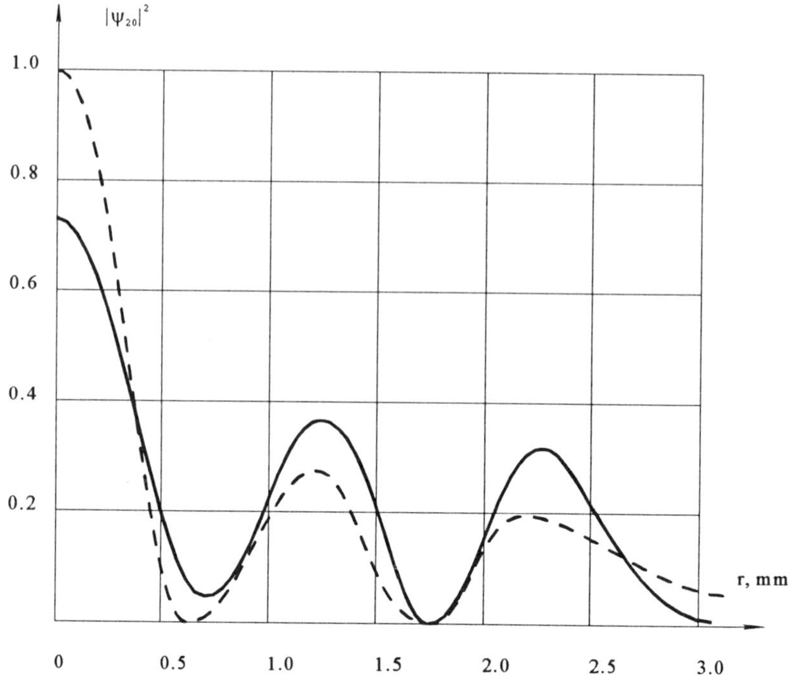


**Figure 5.12.** Intensity distribution in the first diffraction order from a binary-phase modan. r-polar radius measured from the center of mode. (Modified from Bereznyi A.E., Karpeev S.V., Uspleniev G.V., Computer generated holographic optical elements produced by photolithography, *Optics and Lasers in Engineering*, 15(5), 331, 1991).

## 5.3. CORRELATIVE INVESTIGATION OF TWO-BEAM MODANS

In order to prove an operating ability of multibeam modans, a twobeam single modal modan was generated and aimed at simultaneous selecting $L$=2 modes of Gauss-Laguerre with the indices $(p,l)$=(0,0) and $(p,l)$=(1,0), where $\nu$=10 lines/mm. Two singlebeam singlemodal modans characterized by corresponding spatial frequencies were generated simultaneously .

The experimental studies were conducted with the help of an experimental set-up dedicated to the analysis of the transverse-mode composition (see Fig. 5.13). The optical scheme is based on a coherent-optical correlator.

Laser $L$ beam after the expander $P$ illuminates a single-beam modan $M_1$. A filtration system containing two objectives $O_1$ and $O_2$ with equal focal length $f_0$, and a diaphragm $D$ located in the focus of the objective $O_1$, discriminates the image in the first diffraction order.

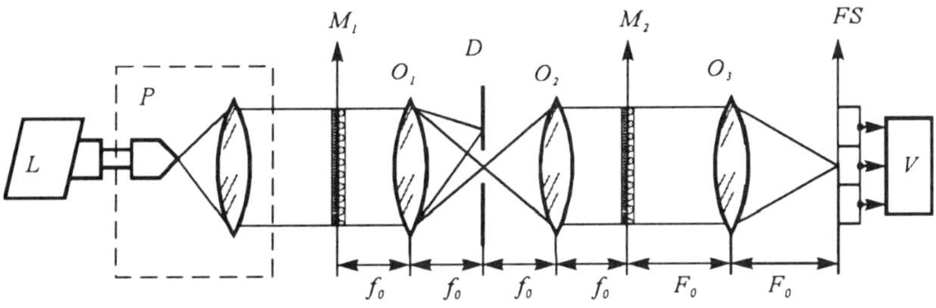


**Figure 5.13.** Optical set-up for the experimental investigation of a multibeam modan. *L*-laser; *P*- beam-expander with pinhole; $O_1$, $O_2$, $O_3$-objectives, $M_1$, $M_2$-modans, *D*-diaphragm in Fourier plane, *FS*-array of photoreceivers; *V*-electric signal indicator; (Modified from Adzhalov V.I., Golub M.A., Karpeev S.V., Sisakyan I.N., Soifer V.A., Multi-channel computer optics components matched to mode groups, *Soviet Journal of Quantum Electronics*, 20(2), 136, 1990).

The formed beam goes through a twobeam modan $M_2$ that performs the separation of transverse modes. An objective *O* is placed in the Fourier-stage of an optical scheme. An array of photoreceivers *FS* is placed at the output plane of the Fourier stage. The focal length of the objective $O_3$ was chosen $F_0$=750 mm. The height of the correlation peak ranged from 100 to 150 μm that corresponded to 8 to 12 cells of a linear photoreceiver array. The results are clearly seen from Fig. 5.14. If we place into a correlator the twobeam modan together with the singlebeam single modal modan designed for the (0,0) mode, the correlative peak can be observed at the point $\lambda F_0 \mathbf{v}^{(0,1)}$ of a focal plane (Fig. 5.14a), whereas for a modan designed for the (1,0) mode, the correlative peak takes place in the different point corresponding to $\lambda F_0 \mathbf{v}^{(1,0)}$ (Fig.5.14b).

In the case the filter forms the beams of different modes, there is no intensity peak (see Fig. 5.14c). Therefore, a twobeam modan actually performs a simultaneous formation of two modal beams characterized by the modes $\psi_{00}$ and $\psi_{10}$, respectively, meanwhile singlebeam modan forms single-modal beam with the modes $\psi_{00}$ or $\psi_{10}$, respectively.

## 5.4 MEASUREMENTS OF MODAL POWER DISTRIBUTION IN GRADED-INDEX FIBERS

The problem of measuring a modal power distribution in optical waveguide arises in such tasks as the design of the coupling components of fiber-optical communication lines, the studies of fiber performances, the optimization of launching conditions in optical fiber and so on.

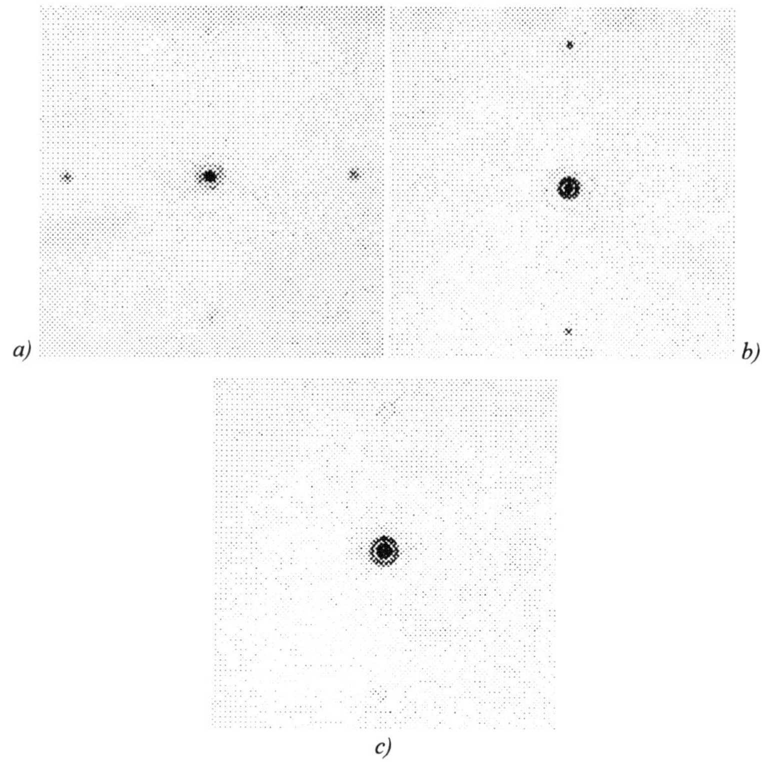

Figure 5.14. Photograph of correlative peaks for twobeam and single-beam modans. (Modified from Adzhalov V.I., Golub M.A., Karpeev S.V., Sisakyan I.N., Soifer V.A., Multi-channel computer optics components matched to mode groups, *Soviet Journal of Quantum Electronics*, 20(2), 136, 1990).

Measuring methods associated with digital processing proposed by [Kitayama, 1979; Shigezawa, 1978] feature high accuracy, but they are difficult in realization and do not allow for data processing in a real time. In section 3.3 of this book we suggested that the analysis of the composition of coherent light transverse modes should be done optically through the use of modans.

This approach makes it possible to perform measurements in a real time and with high accuracy. This section contains the results of experiments on modan-aided measurements of the modal power distribution. For the sake of comparison the modal power distribution is also reconstructed by computer-aided processing of the interferograms of fiber output field .

**A. General-purpose experimental set-up**

In the process of experiments we used the 210 mm length test-segment of a multimodal optical fiber with the parabolic profile of the refractive index described as follows:

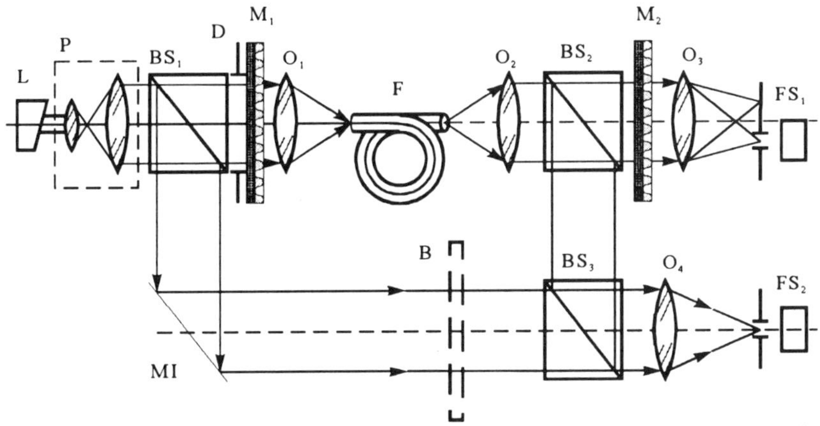


**Figure 5.15.** Optical set-up designed for the selective mode excitation and analysis in an optical fiber. *L* is He-Ne laser; *P*-beam expander with pinhole; $BS_1, BS_2, BS_3$-beam-splitting cubes; *D*-diaphragm; *F*-fiber; $O_1, O_2$-microobjectives; $O_3, O_4$-objectives, $M_1, M_2$-modans, *MI*-mirror, *B*-birefringent plate. (Modified from Golub M.A., Sisakyan I.N., Soifer V.A., Uvarov G.V., New measurement techniques for modal power distribution in fibers, *Proceedings SPIE, OE/Fibers'90*, 1366, 273, 1990).

$$n^2(r) = n_0^2\left[1 - 2\Delta\left(\frac{r}{a}\right)^2\right]$$

with Δ= 0.009, $a$ = 30μm, $n_0$= 1.458, and with the numerical aperture *NA* =0,19. The radius of the principal mode in such a fiber equals $\sigma_0$=5.6 μm. In a variety of experiments performed the same basic optical set-up described in Fig. 5.15 was used and modified. The basic optical set-up for the experimental installation is interferometric and comprises a basic branch with a fiber $F$ ($BS_1 - M_1 - BS_2 - M_2 - FS_1$) and reference branch of the interferometric scheme $BS_1 - MI - BS_3 - FS_2$, both receiving the light from a He-Ne laser $L$ through the collimator $P$ and the beam-splitter $BS_1$.

**B. Modan-aided measurements of modal power distribution in fiber**

For the removed mirror *MI,* an interferogram is not formed, but the reference branch with the photoreceiver $FS_2$ is utilized to perform a relative normalization of the signals from the photoreceiver $FS_1$. In the principal branch the radiation was launched into the fiber $F$ from the modan $M_1$ and the microobjective $O_1$. The microobjective $O_2$ magnifies the output light field from the fiber $F$ end and projects it onto the modan plane $M_2$. It is known from theory (section 3.3) that the modan $M_2$ coupled with the Fourier-objective $O_3$ forms in the $FS_1$ plane the field characterized by an intensity, which is proportional to the modal power (in the points corresponding to the first diffraction order). On the stage of doing an optical modan analysis, the modan $M_1$ was not introduced into set-up, the fiber was excited with a focused beam analogous to the Gaussian one, and the mirror $MI$ was removed.

The singlemodal modans $M_2$ were utilized with the complex transfer $\psi^*_{pl}(\mathbf{x})$ matched to the axially symmetrical Gauss-Laguerre modes numbered $(p,l)$, with $l=0$ and $p=0,1,2,3$.

Thus the modan $M_2$ converts the power distribution between Gauss-Laguerre modes into the intensity distribution in the plane $FS_1$. Intensity measurements have been conducted in the corresponding discrete points with the help of the CCD- array $FS_1$. To perform normalization, the readings of the photo-diode $FS_2$ were employed.

The Table 5.3 shows the modal powers normalized with respect to a principal mode power and measured with the aid of modans.

**C. Digital processing of fiber output field for modal powers restoration**

The modan-aided measuring of the modal power distribution has considerable advantages. However, being a coherent optical method of processing, it is characterized by the error of 10% - 15%. As [Shigezawa, 1978] have shown, a computer method based on the analysis of interferogram of the fiber waveguide field offers more accurate results, though at the expense of the increased execution time. The computer method was implemented (see [Arefyev, 1988]) with the aim of verifying of an optical approach. An optical scheme for the development of the field interferograms from the fiber waveguide $F$, is shown in Fig. 5.15. This is a Mach-Zehnder interferometer. After the expander $P$, a collimated laser beam goes to the beam-splitter $BS_1$ and is divided into the object and reference waves. On passing the iris diaphragm $D$, the object wave is focused by the objective $O_1$ onto the end of the fiber $F$ under study. Such a method for excitation a fiber has been described by [Bartelt, 1983] and allows the exciting beam parameters to be readily altered.

The output distribution is collimated by the objective $O_2$ and then superimposed with the reference beam using the mirror $MI$ and the beam-splitters $BS_2$ and $BS_3$. The reference beam goes through a birefringent plate $B$, the latter allowing for introducing the phase delay $\Delta\varphi$, which is necessary for quadrature interferogram.

The resulting interferometric pattern is projected, through the use of the objective $O_4$, onto the registration plane with the photoreceiver $FS_2$ for a photo camera. On shutting the reference or the object beam, we are able to register each of them separately, that turns out to be necessary for the subsequent processing of the obtained interferograms.

**Table 5.3**

**Modal powers $\left|w_{p0}\right|^2$ versus radial mode number $p$**

| $p$ | 0 | 1 | 2 | 3 |
|---|---|---|---|---|
| $\left\|w_{p0}\right\|^2 / \left\|w_{00}\right\|^2$ | 1 | 0.90 | 0.27 | 0.50 |

Modified by Arefyev E.Yu., Bambulevich K.E., Golub M.A., Karpeev S.V., Sisakyan I.N., Soifer V.A., Comparison of two methods for the determination of the power distribution between modes, *Soviet Journal of Quantum Electronics*, 18(12), 1548, 1988 .

When varying a reference beam tilt, from the known direction of a mirror revolution and from the variation of the interferogram carrier we can determine the direction of the interferogram phase increase. The focal lengths of the objectives $O_1$ and $O_2$ were taken as 50 mm. To enter interferograms into a computer, they were recorded on a photofilm and then imaged by a computer-controlled TV-camera with a frame-grabber. Such an input procedure enables us to govern the transfer characteristic by way of suitable selection of the photo registration regime.

The input channel for image digitizing into the computer was calibrated in several steps. In the first step, the minimum and maximum values of the registered distribution were measured with the aid of a CCD-array in the plane of registration.

A second step used a test photometrical edge that gave us the characteristic of the input unit shown in Fig. 5.16. An optical density of the optical edge is plotted on the x-axis, and the digital readings from the input unit are plotted on the y-axis. The curve has well-defined linear part. From the known range of the exposure intensities and from the required range of optical densities on a photofilm, we have performed suitable selection of a photo material as well as the mode of both its exposure and photochemical processing. The best results have been obtained on the underexposure portion of the characteristic curve when using a KH-I-type (Russian catalogues) film. The through characteristic of the entire process of input into computer is shown in Fig. 5.17 and was obtained with the help of a set of neutral light-filters.

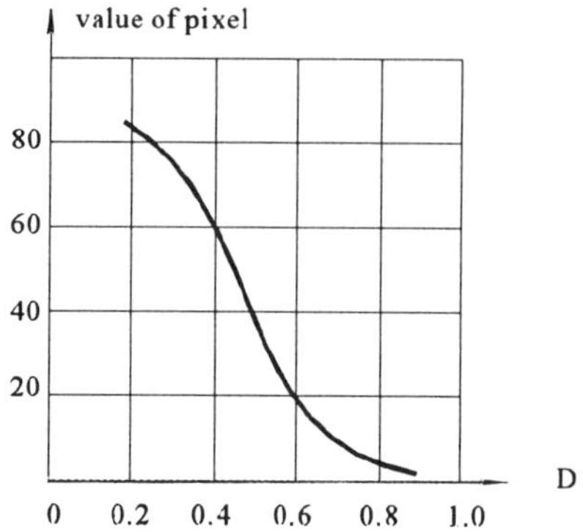


**Figure 5.16.** Input nonlinearity of TV camera measured with the aid of the photometrical edge (Modified by Arefyev E.Yu., Bambulevich K.E., Golub M.A., Karpeev S.V., Sisakyan I.N., Soifer V.A., Comparison of two methods for the determination of the power distribution between modes, *Soviet Journal of Quantum Electronics*, 18(12), 1548, 1988).

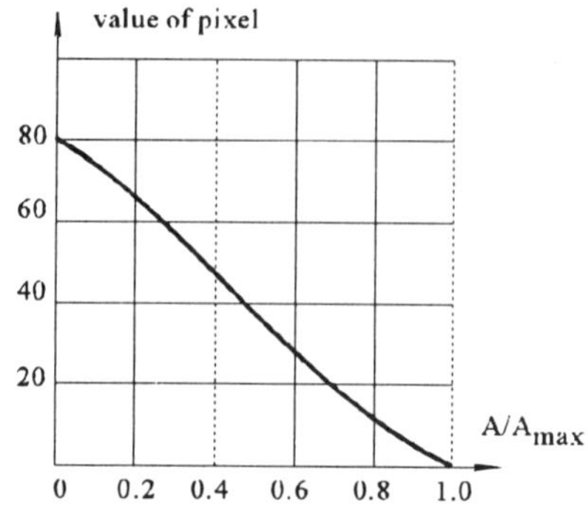


**Figure 5.17.** Nonlinear characteristic of TV camera input of the light field amplitude into the computer (Modified by Arefyev E.Yu., Bambulevich K.E., Golub M.A., Karpeev S.V., Sisakyan I.N., Soifer V.A., Comparison of two methods for the determination of the power distribution between modes, *Soviet Journal of Quantum Electronics*, 18(12), 1548, 1988).

The ratio of the light amplitude $A$ to the maximum amplitude $A_{max}$ is marked on the $x$-axis, and the corresponding digital pixels are marked on the y-axis. As is seen from Fig. 5.17, we managed to obtain a curve characterizing the process of linear input of light amplitude into a computer. Figure 5.18 depicts the obtained interferogram. The object wave intensity considerably varies and becomes zero at some points. This causes the wavefront dislocations (see [Baranova, 1981]) and complicates the selection of reference lines. The decoding of such interferograms was performed in the interactive mode and the stages of processing were displayed on a computer screen.

Prior to entering into the computer, each interferogram was subject to the preliminary processing. That was done to eliminate the pulse noise distorting the separate points or even the entire strings of the input image. Both the nature of input images and the aim of their subsequent processing make it possible to employ a simple median filter (see [Pratt, 1982]) to filtrate the pulse noise. Also, the stage of the preliminary processing involves the finishing correction for the non-linearity as well as the reversal of negative image to a positive one. Let us describe the algorithm of digital processing.

For the aim of computer-aided calculation, the field $w(\mathbf{x})$ and modal functions $\psi_{\mathrm{p}}(\mathbf{x})$ are specified as the matrixes of the discrete pixels $\left[w_{\mathrm{j}}\right]$ and $\left[\psi_{\mathrm{j}}\right]$, with

$$\mathbf{j} = (j_1, j_2);\ j_1 = 1,2,\ldots,N_1;\ \ j_2 = 1,2,\ldots,N_2$$

The purpose is to measure the complex values $w_{\mathrm{j}} = A_{\mathrm{j}}\left(\cos\varphi_{\mathrm{j}} + \sin\varphi_{\mathrm{j}}\right)$. We have recorded two quadrature interferograms and, separately, the intensity fields $A_{\mathrm{j}}^2$ and $B_{\mathrm{j}}^2$ for the object and reference beams. The intensity of the principal interferogram could be written as follows

$$I_{\mathrm{j}} = A_{\mathrm{j}}^2 + B_{\mathrm{j}}^2 + 2A_{\mathrm{j}}B_{\mathrm{j}}\cos\varphi_{\mathrm{j}}, \tag{5.1}$$

The matrices $\left[I_{\mathrm{j}}\right]$ and $\left[A_{\mathrm{j}}\right]$, $\left[B_{\mathrm{j}}\right]$ were measured separately. Then the phase of the reference beam was changed by $\pi/2$ and the field intensity $\left[I_{\mathrm{j}}'\right]$ that included $\sin\varphi_{\mathrm{j}}$ instead of $\cos\varphi_{\mathrm{j}}$ was recorded. Reconstructing the real and the imaginary part of $w_{\mathrm{j}}$, we find from Eq.(5.1) that

$$w_{\mathrm{j}} = A_{\mathrm{j}}\left(\frac{I_{\mathrm{j}} - A_{\mathrm{j}}^2 - B_{\mathrm{j}}^2}{2A_{\mathrm{j}}B_{\mathrm{j}}} + i\frac{I_{\mathrm{j}}' - A_{\mathrm{j}}^2 - B_{\mathrm{j}}^2}{2A_{\mathrm{j}}B_{\mathrm{j}}}\right). \tag{5.2}$$

As compared with the results of [Shigezawa, 1978], the field in Eq.(5.2) can be restored in a more general form, when the reference beam phase is changed by an arbitrary value $\Delta\varphi$.

In this case, to reconstruct the imaginary part $w_{\mathrm{j}}$, the following relationship is used:

$$\sin\varphi_{\mathrm{j}} = \frac{\cos\left(\varphi_{\mathrm{j}} + \Delta\varphi\right) - \cos\varphi_{\mathrm{j}}\cos\Delta\varphi}{\sin\Delta\varphi}$$

Both the model specified by Eq.(5.1) and the decoding method given above are operable only for minor nonlinearities in the interferogram measurements. Such method was applied to the interferogram in Fig. 5.18.

The methods using the selection of reference lines and the interpolation of the phase values between lines proved to be most stable to the nonlinearities of measurements. A selection of reference lines for the interferogram shown in Fig. 5.18 was performed by hard-clipping of intensity with respect to some average level that was chosen during the computer processing.

Figure 5.19 shows the reference lines for interferogram of Fig 5.18. The phase values between lines were found through the linear interpolation. After reconstructing the complex pixels of the field $w_{\mathrm{j}}$ and calculating the modal function $\psi_{\mathrm{j}}$, the mode coefficient $\mathrm{w}_{\mathrm{p}}$ Eq.(3.5) was estimated as follows:

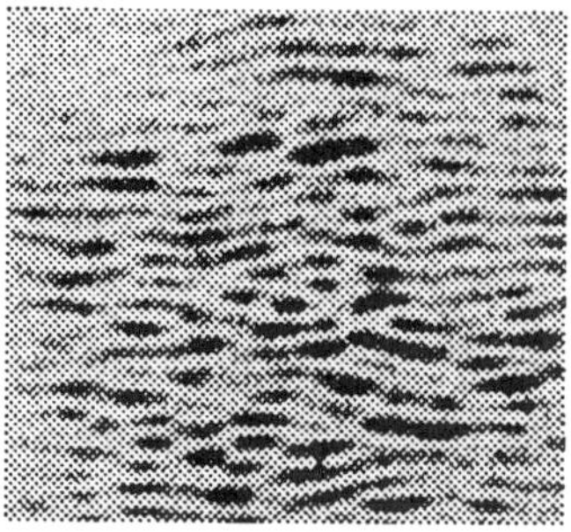

**Figure 5.18.** Interferogram of field at the output end of a graded-index fiber (Modified by Arefyev E.Yu., Bambulevich K.E., Golub M.A., Karpeev S.V., Sisakyan I.N., Soifer V.A., Comparison of two methods for the determination of the power distribution between modes, *Soviet Journal of Quantum Electronics*, 18(12), 1548, 1988).

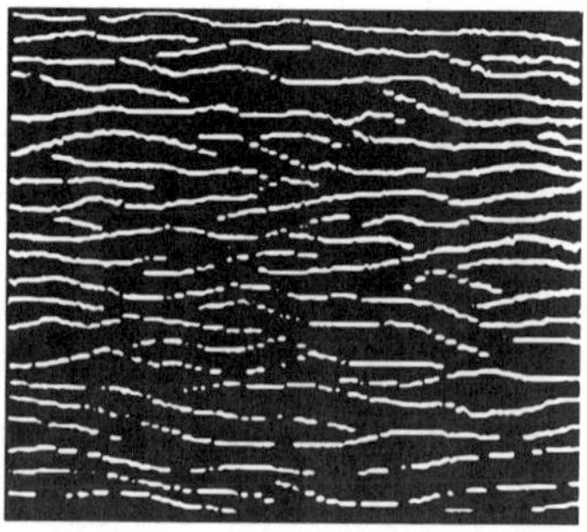

**Figure 5.19.** Reference lines for the interferogram of Fig. 5.18. (Modified from Arefyev E.Yu., Bambulevich K.E., Golub M.A., Karpeev S.V., Sisakyan I.N., Soifer V.A., Comparison of two methods for the determination of the power distribution between modes, *Soviet Journal of Quantum Electronics*, 18(12), 1548, 1988).

$$\hat{w}_{\mathrm{p}} = \frac{1}{N_1 N_2} \sum_{j_1=1}^{N_1} \sum_{j_2=1}^{N_2} w_{\mathrm{j}} \psi_{\mathrm{j}}^{*}$$

and the modal power $\sim \left| w_{\mathrm{p}} \right|^2$. The reconstruction of each field requires large execution time of a computer. Since this experiment was rather time-consuming, we calculated the powers for only four modes numbered (0,0), (0,1), (0,2), and (0,3). Two realizations based on the quadrature interferogram approach and one realization using the reference lines' method were processed.

The results of experimental data processing are presented in Table 5.4. For the sake of comparison, the table includes the results of optical measurements taken from the table 5.3. "Optical" result differs from the results of digital processing by 20 per cent, that proves the operating capability of a modal analyzer employing the computer-generated elements - modans. Alternating large and small values for the modal powers in Table 5.4 with even or odd indices agree with the theory.

The analysis of errors typical of computer-aided method of measuring the modal power distribution showed that the reasons for their emergence are as follows:

1. Auto-regulation of a TV-camera amplification
2. Non-uniformity of the TV-camera characteristics with in the field
3. Small number of pixels in the sampled images.

**Table 5.4**

**Values of modal power distribution $\left|w_{p0}\right|^2 / \left|w_{00}\right|^2$**

| *p* | **0** | **1** | **2** | **3** |
|---|---|---|---|---|
| Digital processing of quadrature interferograms | 1 | 0.54 | 0.11 | 0.46 |
| Digital processing based on reference lines | 1 | 0.90 | 0.27 | 0.50 |
| Optical measurements with the aid of modans | 1 | 0.71 | 0.47 | 0.41 |

Thus, although the computer-aided method offers better accuracy than optical methods, it appears to be rather time-consuming. An essential advantage of optical methods with modans is a one-step measurement procedure. However, the development of measuring facilities will allow one to pick up the speed of the computerized method in the future. At the same time, an essential advantage of the computerized method is its flexibility. The disadvantage an optical method is the worse reproducibility of results as well as the difficulty in adjustment of filters.

Upon detailed analysis, we concluded that the optical method with modans is more suitable for the tasks where one needs to perform on-site of measurements of modal powers, for example, in the case of the technological checking of fibers. At the same time, the computer-aided method proves to be more appropriate for one-time high-precision measurements.

## 5.5 MODE-SELECTIVE EXCITATION OF THE FIBER WITH THE AID OF MODANS

One series of experiments (see [Golub, 1984]) was devoted to the studies of launching defocusing based on the method described by [Bartelt, 1983]. The scheme of Fig. 5.15 was employed with the removed mirror *MI* and modan.

The iris diaphragm $D$ of a varying diameter $d_0$was used. [Bartelt, 1983] have shown that the diffraction pattern at the input end plane of the fiber $F$ offers sufficient approximation of the Gaussian intensity with the beam-waist radius of $\sigma = 1.121\,\lambda f_1/\pi r$, where $\lambda$ is the wavelength of the laser $L$, $f_1$-focal length of $O_1$ objective.

The modan $M_2$ matched to the Gauss-Laguerre modes $(p,l)$, at $l$ =0 and $p = \overline{0,3}$, was manufactured in the form of two spatial filters designed separately for the positive ($\psi^+_{pl}$) and the negative ($\psi^-_{pl}$) parts of the real function.

To produce the fiber excitation, the light beam from a He-Ne laser was expanded and then limited by the iris diaphragm having regulation range of 0.75 to 8 mm. The focal length of the objective $O_1$ was chosen equal to 50 mm. The objective $O_3$, the focal distance 135 mm, provided such a scale in the Fourier plane $FS_1$ that the waist radius for the Gaussian beam is $\sigma_1$=45 μ m. Here $\sigma_1$=1.5 μm.

Since the variations should occur in a point, it is easily seen that the aperture of a photoreceiver should be 3- 5 times less than the radius of $\sigma_1$. For the purpose of the present experiment, a CCD- array was used with the cell size of 15x15 μm. The use of the CCD-array makes it possible to measure the intensity distribution in different cross-sections of the *FS* plane and to align the filter relative to the beam under study. Due to the radial symmetry of the pattern, a single cross section passing through the origin of coordinates contains all the required information. The change of filters was carried out in a mechanical manner through the use of a cassette with the precision alignment.

In the experiment under discussion, powers of the four modes were measured as the functions of the radius $\sigma_1$ of the exciting Gaussian beam in the absence of the beam shift and the tilt relative to the fiber axis. The results of measurements are presented as the solid line in Fig. 5.20 versus the normalized parameter $v = \sigma_0 / \sigma_1$. [Grau, 1980] have derived the following relationship:

$$\frac{|w_{20}|^2}{|w_{00}|^2} = \frac{4v^2}{\left(1+v^2\right)^2}\left[\frac{\left(1-v^2\right)}{\left(1+v^2\right)}\right]^{2p} \tag{5.3}$$

The corresponding theoretical curves are also displayed in Fig. 5.20 as the dotted lines. As is seen from Fig. 5.20, the investigators failed to excite a single principal mode even for $v$ =1. The fact of excitation of the higher order modes in this case can be ascribed to the imperfection of a fiber as well as to the difference of the exciting beam from the Gaussian.

However, the trend of the function $v_{00}$ versus $v$ is in a good agreement with the theory. For the modes $\psi_{20}$ and $\psi_{30}$, the error proves to be somewhat larger, especially for $v$>1.5. This may result from the necessity of obtaining the radius of exciting beam equal about to 1.5 μm, that increases the influence of the aberration produced by the objective $O_1$. One should also note that when measuring modes of higher orders the level of the measured signal becomes very low and comparable to measurement errors of a coherent optical system.

In another series of experiments, the basic interferometer set-up shown in Fig. 5.15 was employed to produce the mode-selective excitation of a fiber. The Gauss-Laguerre modans $M_l$ numbered as $(p,l)$, where $l$ = 0, and $p$ =0,1,2 were placed in turn at the principal branch input. If the reference branch is shut, the field intensity excited in the fiber $F$ is recorded. The photograph in Fig. 5.21 shows the far field of the fiber output end under the excitation by the modes with the indices $(p,l)$ = (0,0) and (1,0) generated by the modans. The selective excitation of the said modes is observed.

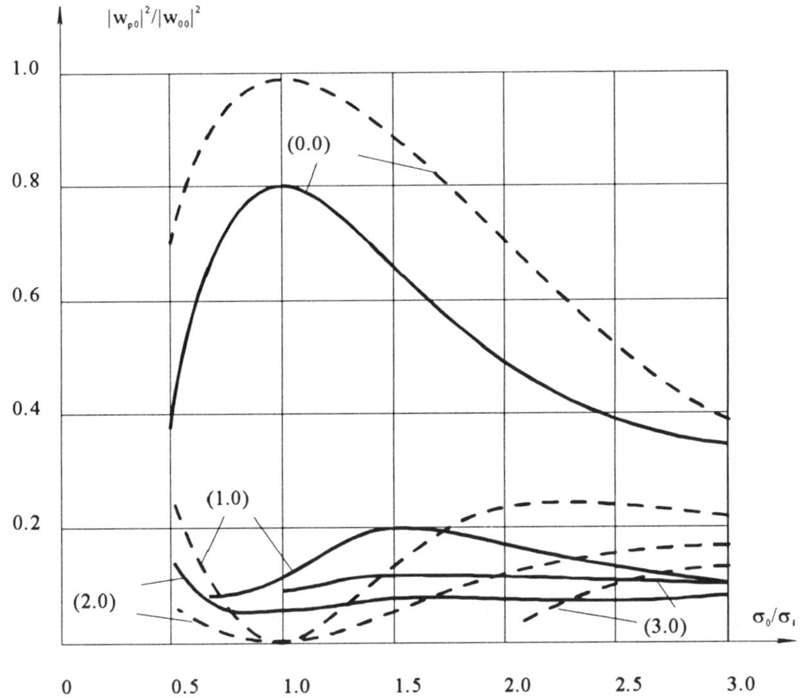


**Figure 5.20.** Modal power variations due to defocusing. $\sigma_1$-radius of launching beam, $\sigma_0$-waist radius of principal mode in the fiber. (Modified from Golub M.A., Karpeev S.V., Krivoshlykov S.G., Prokhorov A.M., Sisakyan I.N., Soifer V.A., Spatial filter investigation of the distribution of power between transverse modes in a fiber waveguide, *Soviet Journal of Quantum Electronics*, 14(9), 1255, 1984).

For the purpose of interferometric studies, the portion of an input radiation after passing through the beam-splitting cube $BS_1$ and the mirror $MI$ is conveyed (as a reference beam) to the beam-splitting cube $BS_3$. It is behind $BS_3$ cube that the interferogram of the fiber's output field is formed. Figure 5.22 depicts the interferograms for the far field at the fiber output. Computer -generated pictures of the same modes are presented on the Fig. 5.23 for the sake of comparison.

Therefore, the developed modans make it possible to produce in a graded-index fiber the selective excitation of the Gauss-Laguerre modes characterized by various indices.

## 5.6 TOLERANCES FOR LAUNCHING CONDITIONS IN A MODAN-AIDED EXCITATION OF THE FIBER

The waist radius of the principal mode is relatively small for really utilized optical fibers, (equal to 5.6 μm for our fiber).

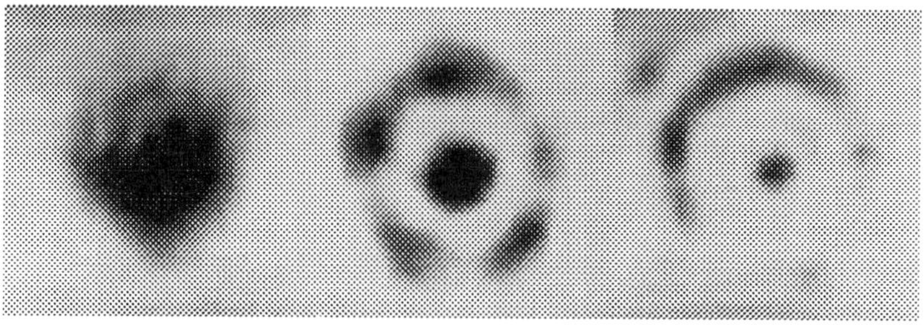

**Figure 5.21.** Photograph of the Gauss-Laguerre modes excited in a fiber with the aid of modans. (Modified from Garitchev V.P., Golub M.A., Karpeev S.V., Krivoshlykov S.G., Sisakyan I.N., Soifer V.A., Uvarov G.V., Use of synthesized holograms for selective mode excitation in gradient index fibers, Analysis of sensitivity to radial displacement of launch beams, in *Computer Optics*, Pergamon Press, London, 2(1), 95, 1990).

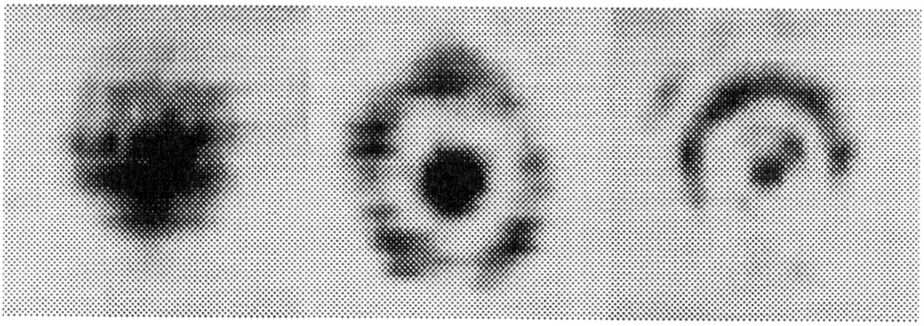

**Figure 5.22.** Interferograms of the Gauss-Laguerre modes excited in the fiber with the aid of modans. (Modified from Garitchev V.P., Golub M.A., Karpeev S.V., Krivoshlykov S.G., Sisakyan I.N., Soifer V.A., Uvarov G.V., Use of synthesized holograms for selective mode excitation in gradient index fibers, Analysis of sensitivity to radial displacement of launch beams, in *Computer Optics*, Pergamon Press, London, 2(1), 95, 1990).

This, in fact, imposes stringent requircments on the adjustment unit for the mode-selective excitation of a fiber. [Karpeev, 1988] have conducted investigation on how launching conditions affect the excitation of a single mode. With the aim of determining the tolerances, it has been studied (see [Golub, 1990]) how the modal power for selectively excited modes depends on (1) the degree of decentering, (2) the variations in the scale or in the tilt.

The optical set-up shown in Fig. 5.15 only, with the reference branch absent, was employed. To provide the singlemodal operation of a fiber, input the coherent field at the fiber was formed by the suitable modans $M_1$ matched to the appropriate axially symmetrical Gauss-Laguerre modes numbered $(p,l)$, where $l = 0$ and $p = 0,1,2$. The modans $M_1$ and $M_2$ were employed both at the fiber's input and at the output.

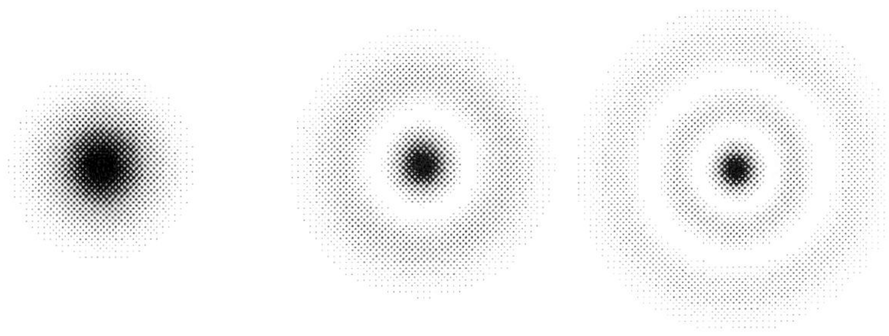

**Figure 5.23.** Pictures of the ideal Gauss-Laguerre modes numbered as $(p,l)$: (0,0), (1,0) and (2,0).

Thus, the variations in the power of the particular mode could be measured that was formed in the fiber under the different physical launching conditions. A special technique was used for the initial adjustment of a modal analyzer with respect to the fiber. For this modan designed for the modes with the index $(p',0)$ different from $(p,0)$ was used as $M_2$. Thus, the dark spot with the light halo was provided in a focal plane. On the contrary, a modan corresponding to the exciting mode $(p,0)$ was placed if at the $F$ fiber output, than a rather pronounced correlative peak occurring at the photoreceiver's aperture. Below we discuss the $(p,0)$ mode power measurements taken at the output of the fiber $F$, with the same exciting mode $(p,0)$ launched at the input. The following parameters of launching conditions were varied: the decentering, the tilt and the defocusing that were respectively specified by the parallel shift $\Delta r$, the mutual tilt $\theta$ of the fiber axes, with respect to the exciting beam, and the mismatch of the formed mode parameter $\sigma_1$ relative to the principal mode radius $\sigma_0$ of the fiber waveguide.

The solid lines at the Figs. 5.24 through 5.26, depict experimental data on variations of the $(p,0)$ modes' power, whereas the dotted lines represent corresponding theoretical characteristics derived from the relationships for specifying each of the above parameters separately (see [Grau, 1980; Krivoshlykov, 1983]). It is seen from the plots in Fig. 5.24 that the tolerable intensity decreases up to the 0.9 level allows the acceptable radial decentering not to exceed 1.5 μm for the principal mode (0,0), 1.1μm - for the mode (1,0), and 0.8 μm - for the mode (2,0). Such an accuracy of centering could be only achievable by the individual adjustment of each mode.

The studies of the curves in Fig. 5.26 have shown that the focal length of the $O_3$ objective that provides the necessary size of the exciting beam, should lie within the definite tolerance limits. These focal tolerances equal 14- mm for the (0,0) mode, 16-22 mm for the (1,0) mode, and 17-20 mm for the mode (2,0), given the tolerable decrease of the intensity is equal to 0.9.

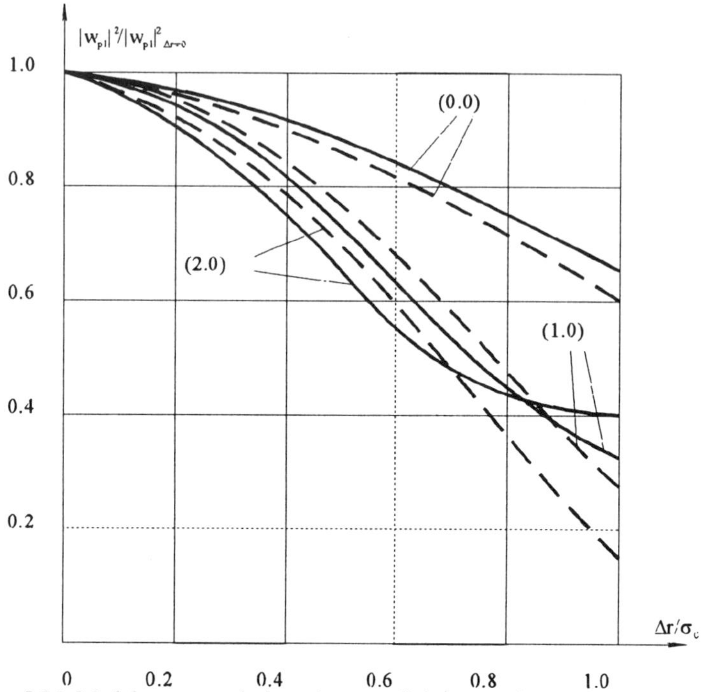


**Figure 5.24.** Modal power variations due to radial decentering $\Delta r$ of launching beam center with respect to the fiber's center. (Modified from Golub M.A., Sisakyan I.N., Soifer V.A., Uvarov G.V., New measurement techniques for modal power distribution in fibers, *Proceedings SPIE, OE/Fibers'90*, 1366, 273, 1990).

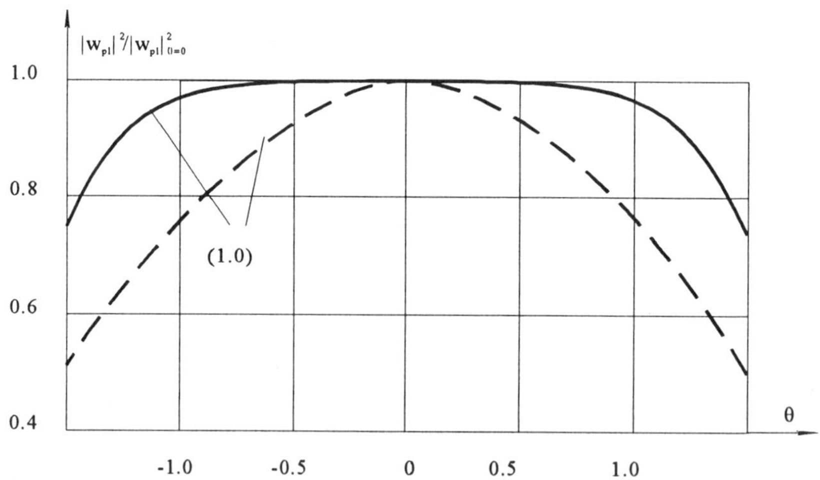


**Figure 5.25.** Modal power variations due to the $\theta$-angle tilt between fiber's axis and launching beam axis. (Modified from Golub M.A., Sisakyan I.N., Soifer V.A., Uvarov G.V., New measurement techniques for modal power distribution in fibers, *Proceedings SPIE, OE/Fibers'90*, 1366, 273, 1990).

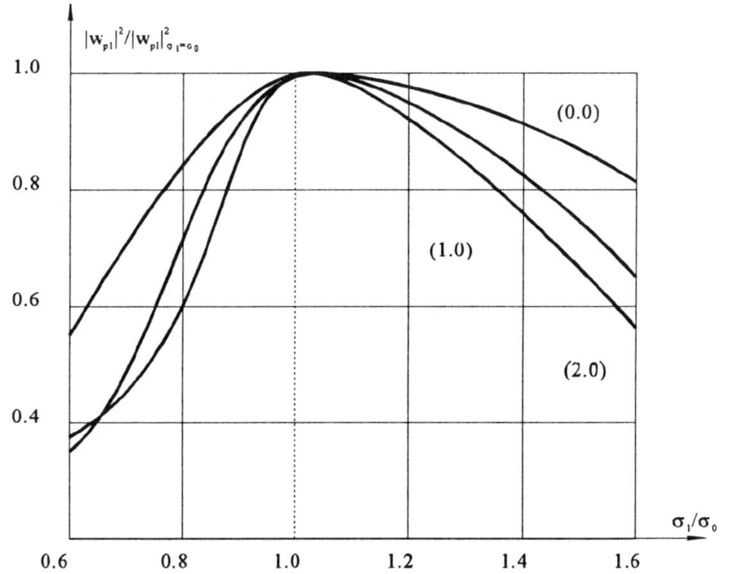


**Figure 5.26.** Modal power distributions upon defocusing $\sigma_1$-radius of illuminating beam, $\sigma_0$-waist radius of principle mode in the fiber. (Modified from Golub M.A., Sisakyan I.N., Soifer V.A., Uvarov G.V., New measurement techniques for modal power distribution in fibers, *Proceedings SPIE, OE/Fibers'90*, 1366, 273, 1990).

Therefore, the tolerance limits become more stringent with the increasing mode order. The experimental investigations reported substantiate the choice of the basic design parameters of optical adjustment units for fiber mode excitation with modans.

# Chapter 6

# FIBEROPTIC SENSORS AND COMMUNICATIONS WITH MODE SELECTION

## 6.1. FIBEROPTIC COMMUNICATION LINE WITH MODAL MULTIPLEXING

The transverse modes' selection offers new opportunities in terms of solving various applied tasks. One should especially note an opportunity to do multiplexing in fiberoptic communication lines through the selective use of transverse modes. This necessitates the formation of a multimodal beam at the transmitting end of channel. Independent modulation of the power for each transverse mode must be available at the transmitting end of channel. In this case the selection and demodulation of each transverse mode take place at the receiving end. As an engineering matter, this problem only awaits its hardware realization, whereas the methods for forming and analyzing the modal composition are entirely prepared for industrial application as described in previous chapters of this book.

It is conventional for fiberoptic communication lines that the information is transferred by the physical value of light flux in a fiber, which is the total power of all the guided modes (see [Cozannet, 1981]). But employing the set of modes with different propagation constants yields us the pulse blurring with the decreasing optical fiber channel capacity.

Entirely new physical approach to the problem of building the fiberoptic communication lines, is proposed here. The approach is based on modans as optical spatial filters, which perform selective formation and separation of transverse modes in an optical fiber. The use of the modans results in the essential improvement of the fiber channel capacity since the data transfer occurs in a parallel fashion through several mode branches, with no pulse blurring observed in each branch. In this way one can obtain the modal channel multiplexing together with the common-used frequency multiplexing.

One of the versions of the developed fiberoptic communication lines is shown in Fig. 6.1. The spatial filter MA represents the array of micro-cells of the "signal-to-light" type modulators illuminated by the plane wave of coherent light.

The cells are connected directly to the source time signals $\{S_i$ the latter being time-modulated and -coded using conventional data transfer methods described in the book of [Gadliardi, 1976]. The Fourier-objective $O_1$ transforms the set of the point sources into the set of the plane waves striking the modan $M_1$.

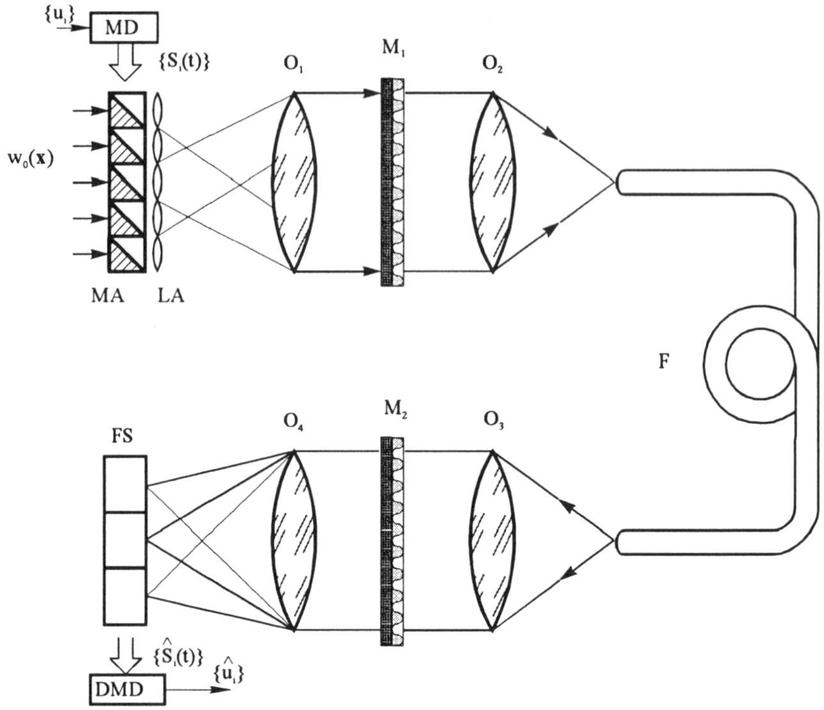


**Figure 6.1.** The scheme of fiberoptic communication line with modal multiplexing; $w_0(\mathbf{x})$-illuminating beam; $\{u_i(t)\}$-vector input message; $\{S_i(t)\}$-temporally modulated vector signal; *MD* is modulator of electrical signal, *MA*-array of optical modulator cells (acoustic-optical or electro-optical); *LA*- lenslet array; $O_1, O_2, O_3, O_4$-objectives; $M_1, M_2$- multibeam modans; *F*-optical fiber; *FS*-photoreceiver array; *DMD*-demodulator of electrical signal; $\{\hat{S}_i(t)\}$-temporally demodulated electrical vector signal; $\{\hat{u}_i(t)\}$-output vector message.

The modan $M_1$ is spatial filter matched to the set of modes recorded at the multiple spatial frequencies and operates using the algorithm of the inverse mode transform (see section 3.4). The carrier spatial frequencies on modan $M_1$ are matched with the coordinates of cells of modulators and their distance from objective $O_1$ At the modan $M_1$ output, there is a superposition of modes characterized by powers proportional to the intensity of the light transmitted through the corresponding cells of modulator $MA$. The microobjective $O_2$ matches the scale of modes to that of the modes in the $F$ fiber cross-section.

A multimodal beam launched into the fiber $F$ is analyzed at the output of the optical fiber communication line. The microobjective $O_3$ matches the modes of the fiber $F$ to the scale of modes of the modan $M_2$, the latter performing the direct modal transform (see section 3.4). The modan $M_2$ is matched to the complex-conjugated modes registered on the multiple spatial frequencies. The light intensity on the microcells of the photoreceiver $FS$ is proportional to the power of the corresponding modes.

One should note that because of the different phase velocities of separate modes their powers are measured with the time delays which are proportional to the propagation constants. To compensate for the differential modal delays and attenuation through the computation, it turns out expedient to choose the "compensating" phases in the branches of the modan $M_2$ based on the equation (see section 2.2):

$$\alpha^{(\mathbf{p})} = \left(\beta_{\max} - \beta_{\mathbf{p}}\right) z$$

where $\beta_{\mathbf{p}}$ is the propagation constant for the **p**-th mode,

$$\beta_{\max} = \max_{\mathbf{p}} \beta_{\mathbf{p}} \ ,$$

and z is the fibers' length. The values of the power distribution between branches which is specified in the process of calculation of modans, are expedient to be chosen as follows

$$\kappa^{(\mathbf{p})} = \frac{1}{\rho_{\mathbf{p}}}, \qquad (6.1)$$

where $\rho_{\mathbf{p}}$ is the modal differential attenuation.

If noises in an optical fiber channel taken into account, it appears expedient to introduce into Eq.(6.1) the filtration term in the form

$$\kappa^{(\mathbf{p})} = \frac{1}{\rho_{\mathbf{p}}} \cdot \frac{1}{1 + \dfrac{S_{\mathbf{p}}}{N_{\mathbf{p}}}},$$

where $S_{\mathbf{p}}$ is the average expected power of a signal for the **p**-th mode at the input of the branch, and $N_{\mathbf{p}}$ is the average expected power of noise for the same **p**-th mode.

### 6.2. MODE-SELECTIVE FIBER-OPTICS SENSORS

Conventional fiberoptic sensors operate on the basis of measuring a total light flux. At the same time, the light flux includes the powers of all modes in a fiber. Since modal powers are summed up and averaged in their variation in the operating sensor, the light flux varies slowly as compared with the power of modal components.

The authors with colleagues have developed an alternative approach and proposed that the modal powers should be differentially detected in the process of measuring the physical value under investigation [Garitchev, 1985; Golub, 1991]. Study of the literature devoted to the mode coupling in fiber waveguides (see [Grau, 1980; Krivoshlykov, 1983] [ Krivoshlykov, 1987]) makes it possible to confirm the idea that the sensitivity of separate modes' power is much greater than that of the light flux.

A novel "mode-selective" fiberoptic sensor can be designed on the basis of such an approach. The main difficulty on this way is to ensure an optical selective excitation of modes at the fiber input and selective detection of modes at the fiber output. However, this difficulty can be eliminated through the use of modans as computer synthesized optical elements. In this section we discuss fiberoptic sensors, with the modans to be placed at a fiber input and output.

Let us consider a sensor with a fiber that has guided modes specified by the complex amplitudes $\{\psi_{\mathbf{p}}(\mathbf{x})\}$ in the scalar approximation with $\mathbf{p}=(p,l)$ as the number of the mode. The Cartesian coordinates $(x,y)=\mathbf{x}$ are introduced here in a fiber cross-section,.

The condition, Eq.(1.12), of the mode orthogonality corresponding to the fiber is supposed to be satisfied in the domain $G$ $(\rho_0 \equiv 1)$. The excitation of the fiber yields the launching beam characterized by the complex amplitude

$$w(\mathbf{x}) = \sum_{\mathbf{p}} w_{\mathbf{p}} \psi_{\mathbf{p}}(\mathbf{x}) \tag{6.2}$$

and having modal composition described by the coefficients $\{w_{\mathbf{p}}\}$.

Describing the propagation of a beam in a fiber we shall use the linear operator $\hat{T}$ that operates on the function $w$. The complex amplitude $W(\mathbf{x})$ of the beam at the fiber's output is described by the equation

$$W = \hat{T} w.$$

The modal composition $\{W_{\mathbf{p}'}\}$ of the output beam

$$W(\mathbf{x}) = \sum_{\mathbf{p}'} W_{\mathbf{p}'} \psi_{\mathbf{p}'}(\mathbf{x}) = \sum_{\mathbf{p}} w_{\mathbf{p}} \hat{T} \psi_{\mathbf{p}}(\mathbf{x}) \tag{6.3}$$

is related both to the modal composition $\{w_{\mathbf{p}}\}$ of the launching beam and to the fiber properties specified by the $\hat{T}$ operator. The kind of the operator $\hat{T}$ depends on the fiber's state. That state is dependent on the disturbances acting on fiber that are characterized by parameters $\mathbf{V} = (V_1, V_2, ...)$ under measurement (see Appendix 3). Thus, we find that $\hat{T} = \hat{T}(\mathbf{V})$, $W_{\mathbf{p}} = W_{\mathbf{p}}(\mathbf{V})$.

In the absence of disturbances on the sensor, we have **V**=0, $\hat{T}=\hat{T}(0)$, where $\hat{T}(0)$ is the operator that describes mode's propagation in their native fiber media.

Using the definition of the propagation constant $\beta_{\mathbf{p}}$ we can write down

$$\hat{T}(0)\,\psi_{\mathbf{p}}(\mathbf{x})=\exp\left(i\beta_{\mathbf{p}}z\right)\psi_{\mathbf{p}}(\mathbf{x})\delta_{\mathbf{pp}'}, \tag{6.4}$$

where $z$ is the distance passed by the mode in the fiber $\delta_{\mathbf{pp}'}$. Note that in this case without disturbances each mode propagates without overlap with others, that is, we have

$$W_{\mathbf{p}'}=\exp\left(i\beta_{\mathbf{p}'}z\right)w_{\mathbf{p}'},\quad \left|W_{\mathbf{p}'}\right|^2=\left|w_{\mathbf{p}'}\right|^2, \tag{6.5}$$

This matches to the fact that the modes are eigenfunctions of the operator describing the light propagation in the non-disturbed optical fiber. The values of the **p** indices correspond to the guided modes in the fiber.

In the presence of non-zero disturbances **V** affecting the fiberoptic sensor, instead of Eq.(6.4) we introduce coefficients $\mathbf{T}_{\mathbf{pp}'}$ so that

$$\hat{T}(\mathbf{V})\psi_{\mathbf{p}}(\mathbf{x})=\sum_{\mathbf{p}'}T_{\mathbf{p}'\mathbf{p}}(\mathbf{V})\psi_{\mathbf{p}'}(\mathbf{x}), \tag{6.6}$$

The coefficients $\hat{T}_{\mathbf{pp}'}(\mathbf{V})$ are the matrix elements of the $\hat{T}$ operator and also referred to as overlap the integrals of the modes (see Appendix 3).The orthonormality of the $\psi_{\mathbf{p}}$ modes gives us the following equation:

$$T_{\mathbf{p}'\mathbf{p}}(\mathbf{V})=\int_G\psi^*_{\mathbf{p}'}(\mathbf{x})\,\hat{T}\psi_{\mathbf{p}}(\mathbf{x})d^2\mathbf{x}.$$

The coefficients $W_{\mathbf{p}}$ instead of Eq.(6.5) can be expressed by the equation

$$W_{\mathbf{p}'}(\mathbf{V})=\int_G W(\mathbf{x})\,\psi_{\mathbf{p}'}(\mathbf{x})d^2\mathbf{x} \tag{6.7}$$

$$W_{\mathbf{p}'}(\mathbf{V})=\sum_{\mathbf{p}}T_{\mathbf{p}'\mathbf{p}}(\mathbf{V})w_{\mathbf{p}} \tag{6.8}$$

derived from Eqs.(1.2) and (6.3).

By virtue of the Parseval inequality for orthogonal expansions (see [Rudin, 1973]), the beams' powers at the fiber input and output respectively are specified by the series:

$$E=\sum_{\mathbf{p}}\left|w_{\mathbf{p}}\right|^2,\quad E'(\mathbf{V})=\sum_{\mathbf{p}'}\left|W_{\mathbf{p}'}(\mathbf{V})\right|^2 \tag{6.9}$$

Most of the common fiberoptic sensors operate by measuring the total output power $E'(\mathbf{V})$ versus the perturbations $\mathbf{V}$. By virtue of Eq.(6.9), the value $E'(\mathbf{V})$ averages different variations in the mode powers $\left|w_{\mathbf{p}'}(\mathbf{V})\right|^2$ over all $\mathbf{p}'$ mode-numbers excited.

The proposed sensor performs selective measurements of powers $\left|W_{\mathbf{p}'}(\mathbf{V})\right|^2$ of the separate modes. Note that in a special case single-modal excitation of the fiber the non-zero coefficient $w_{\mathbf{p}} = 1$ occurs only for one fixed value of $\mathbf{p}$. In this case Eqs.(6.8) and (6.9) take the form

$$W_{\mathbf{p}'}(\mathbf{V}) = T_{\mathbf{p}'\mathbf{p}}(\mathbf{V})$$

$$E'(\mathbf{V}) = \sum_{\mathbf{p}'} \left|T_{\mathbf{p}'\mathbf{p}}(\mathbf{V})\right|^2 .$$

The value $\left|T_{\mathbf{p}'\mathbf{p}}(\mathbf{V})\right|^2$ is also called the mode coupling coefficient since it determines the power of the mode with the number $\mathbf{p}'$, for the case that the fiber is excited by the mode numbered $\mathbf{p}$. [Krivoshlykov, 1983] [Krivoshlykov, 1987] have investigated the coupling coefficients for the cases where a fiber was excited by the modes of different orders. [Taylor, 1971; Unger, 1977] have studied how the fiber microbends affect the modal composition.

## 6.3. MODAN PERTURBATION EFFECTS IN SENSOR OPERATION

Quantization and sampling effects in computer-generated modan cause some parasitic modes excited together with required modes (see chapter 4 for detailed analysis).

This is treated as perturbation of modan. Various kinds of other perturbations can take place too. Thus the sensor's output signal versus the measured parameters becomes a bit different.

Let the fiber be excited by the multimodal beam (see section 6.1) and the output signal be measured with the aid of the singlemodal multibeam modan. Perturbations effect can be described as the replacement of the exciting beam $w(\mathbf{x})$ by the perturbed beam

$$\hat{w}(\mathbf{x}) = \sum_{\mathbf{p}} w_{\mathbf{p}} \varphi_{\mathbf{p}}^{(0)}(\mathbf{x}),$$

with the perturbed modan functions to be found from Eq.(4.10) (for the exciting modan). The output beam $\hat{W} = \hat{T}\hat{w}$ has the following form

$$\hat{W}(\mathbf{x}) = \sum_{\mathbf{p}'} w_{\mathbf{p}} \hat{T} \varphi_{\mathbf{p}}^{(0)}(\mathbf{x}),$$

where the (0) index specifies the perturbations, given the zero carrier $v_{\mathbf{p}}^{(\mathbf{k})} = 0$ in all branches. This is made to have all excited modes going together.

Allowing for Eq.(4.109), the coefficients in Eq.(6.7) are replaced by the perturbed coefficients as follows

$$\hat{W}_{\mathbf{p}'}(\mathbf{V}) = \int_G \hat{W}(\mathbf{x})\varphi_{\mathbf{p}'}^{(\mathbf{p}')*}(\mathbf{x})d^2\mathbf{x}.$$

When taking into account Eqs.(4.10) and (6.6) as well as the orthonormality of modes, Eq.(1.2), we find that

$$\hat{W}_{\mathbf{p}'}(\mathbf{V}) = \sum_{\mathbf{p}}\sum_{\mathbf{q}''}\sum_{\mathbf{q}'}\left[\delta_{\mathbf{p}'\mathbf{q}'} + H_{\mathbf{p}'\mathbf{q}'}^{(\mathbf{p}')}\right]\cdot\left[\delta_{\mathbf{p}'\mathbf{q}''} + H_{\mathbf{p}'\mathbf{q}''}^{(0)*}\right]T_{\mathbf{q}'\mathbf{q}''}(\mathbf{V})\,w_{\mathbf{p}}.$$

If we perform transformations allowing for Eq.(6.8) for the non-perturbed coefficients, we shall find that

$$\begin{aligned}\hat{W}_{\mathbf{p}'}(\mathbf{V}) - W_{\mathbf{p}'}(\mathbf{V}) = \sum_{\mathbf{p}}\sum_{\mathbf{q}'} H_{\mathbf{p}'\mathbf{q}'}^{(\mathbf{p}')}\,T_{\mathbf{p}\mathbf{q}'}(\mathbf{V})w_{\mathbf{p}} + \sum_{\mathbf{p}}\sum_{\mathbf{q}''} H_{\mathbf{p}'\mathbf{q}''}^{(0)*}T_{\mathbf{p}'\mathbf{q}''}(\mathbf{V})\,w_{\mathbf{p}} + \\ + \sum_{\mathbf{p}}\sum_{\mathbf{q}'}\sum_{\mathbf{q}''} H_{\mathbf{p}'\mathbf{q}'}^{(\mathbf{p}')}\,H_{\mathbf{p}\mathbf{q}''}^{(0)*}\,T_{\mathbf{q}'\mathbf{q}''}(\mathbf{V})\,w_{\mathbf{p}}\end{aligned} \tag{6.10}$$

Note that the summation over the **p** index only is made according to the number of the modes to be excited at the fiber. The summation over the remaining indices **q'** and **q''** is extended to infinite limits.

As an of example, let us consider an important special case of the singlemodal excitation of the fiber by the mode numbered **p,** that is, we have $w_{\mathbf{p}} \neq 0$ and $w_{\mathbf{p}''} = 0$, for all other coefficients with $\mathbf{p}'' \neq \mathbf{p}$. Equation (6.10) is simplified as follows:

$$\begin{aligned}\left[\hat{W}_{\mathbf{p}'}(\mathbf{V}) - W_{\mathbf{p}'}(\mathbf{V})\right]/w_{\mathbf{p}} = \sum_{\mathbf{q}'} H_{\mathbf{p}'\mathbf{q}'}^{(\mathbf{p}')}\,T_{\mathbf{q}'\mathbf{p}}(\mathbf{V}) + \sum_{\mathbf{q}''} H_{\mathbf{p}\mathbf{q}''}^{(0)*}\,T_{\mathbf{p}'\mathbf{q}''}(\mathbf{V}) + \\ + \sum_{\mathbf{q}'}\sum_{\mathbf{q}'} H_{\mathbf{p}'\mathbf{q}'}^{(\mathbf{p}')}\,H_{\mathbf{p}\mathbf{q}''}^{(0)*}\,T_{\mathbf{q}'\mathbf{q}''}(\mathbf{V})\end{aligned} \tag{6.11}$$

Using the Cauchy-Bunyakovski inequality as applied to Eq.(6.11) and allowing for the designations in Eq.(4.8), we obtain the following estimate:

$$\begin{aligned}\left|\hat{W}_{\mathbf{p}'}(\mathbf{V}) - W_{\mathbf{p}'}(\mathbf{V})\right|/w_{\mathbf{p}} \leq \left[Q_{\mathbf{p}'\mathbf{p}'}^{(\mathbf{p}',\mathbf{p}')}\cdot\sum_{\mathbf{q}'}\left|T_{\mathbf{p}\mathbf{q}'}(\mathbf{V})\right|^2\right]^{1/2} + \\ + \left[Q_{\mathbf{p},\mathbf{p}}^{(0,0)}\cdot\sum_{\mathbf{q}''}\left|T_{\mathbf{p}'\mathbf{q}''}(\mathbf{V})\right|^2\right]^{1/2} + O\left(H^2\right)\end{aligned} \tag{6.12}$$

where $O(H^2)$ is a second-order infinitesimal with respect to perturbations. Inequality (6.12) makes it possible to estimate the sensor's relative error resulting from modans' perturbations.

## 6.4. FIBEROPTIC SENSOR OF PRESSURE WITH REFLECTIVE COMPUTER-GENERATED OPTICAL ELEMENTS

In the sensor developed, the modan M is mathematically integrated with the Fresnel reflective lens on the stage of computer generation. The design of the sensor is shown in Fig. 6.2. Two parallel line segments $F_1$ and $F_2$ of the graded-index optical fiber characterized by the quadratic profile of the refractive index and with the core diameter of $2a$, are placed at the distance $2H$ to each other. The radius of the principal mode in each fiber is designated as $\sigma_0$. The reflecting-type diffractive optical element in the form of the membrane $C$ is located in the plane that is, perpendicular to the axes of the fiber $F_1$ and $F_1$ and spaced at the distance $l$ from their output ends. The light beam coming from the fibers $F_1$ has the waist at the fiber $F_1$ end and works as an illuminating beam for the respective lens M with a modan. The distances and symmetry of optical set-up are chosen so that the reflected beam produces the excitation of the fiber $F_2$. When the membrane is at the stationary position, the light beam reflected from $M$ has the waist at the $F_2$ fiber input end exactly. External disturbances resulting in longitudinal and transverse shifts or oscillations in the membrane mismatch M and $F_2$ and produce the modal power modulation in the fiber $F_2$. If the element $M$ is a reflecting lens, there occurs the modulation of the total light beam.

Let us introduce the Cartesian coordinates $\mathbf{x}=(x,y)$ in the plane of the optical element $M$. Let the phase function of $M$ be as follows

$$\varphi_M(\mathbf{x}) = \varphi_\Lambda(\mathbf{x}) + \varphi(\mathbf{x}), \tag{6.13}$$

where $\varphi_\Lambda(\mathbf{x})$ is the phase function of the lens and $\varphi(\mathbf{x})$ is the phase function of the modan that was computed in sections 2.2 - 2.5. In this case the power of a single mode or a modal group will come to the input of the fiber $F_2$ selected from the fiber's beam $F_1$.

Aiming to determine the phase function $\varphi_\Lambda(\mathbf{x})$, let us consider a diffractive analog of an asymmetric mirror, with the pair of aplanatic points lying in the paraxial points of curvature of the Gaussian beams of the first and the second segments of the fiber. Since the waists of the Gaussian beams are coplanar to the ends of the $F_1$ and $F_2$ fibers, the centers of their curvature $C_1$ and $C_2$ have coordinates $(-H,0,-L)$ and $(H,0,-L)$, respectively (with regard to the center $O$ of the planar element), where

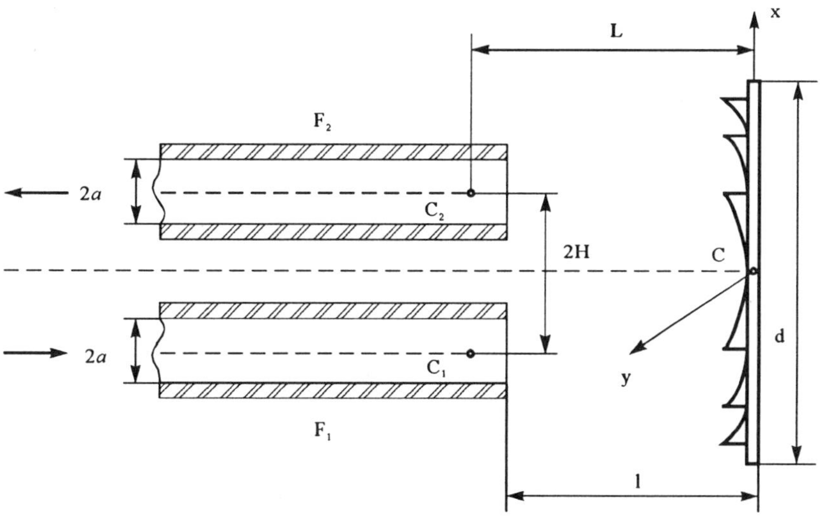


**Figure 6.2.** Fiber sensor of pressure based on the reflecting modan. $F_1$, $F_2$ - input and output fibers; $M$ - reflective-type modan combined with Fresnel lens; $C_1$, $C_2$ - centers of curvatures of the wavefronts of Gaussian beams going from the first fiber and to the second fiber respectively. (Modified from Golub M.A., Kazanskiy N.L., Prokhorov A.M., Sisakyan I.N., Soifer V.A., Synthesis of optical antennae, *Computer Optics*, Pergamon Press, London, 1(1), 25, 1989).

$$L = l\left(1 + \frac{z_0^2}{l^2}\right), \quad z_0 = \frac{k\sigma_0^2}{2}, \quad k = \frac{2\pi}{\lambda}.$$

According to the condition of aplanatism, the spherical wave front diverging from the point $C_1$ should be transformed into the spherical wavefront $\sigma_2$ converging to the point $C_2$. Since the fiber $F_1$ produces in the plane $M$ the field characterized by the phase $k\psi_0$

$$\psi_0(\mathbf{x}) = \sqrt{(x+H)^2 + y^2 + L^2} - L$$

and the beam launching into the fiber $F_2$, should ensure the phase $k\psi$,

$$\psi(\mathbf{x}) = L - \sqrt{(x-H)^2 + y^2 + L^2}\,.$$

Thus the phase function can be found from the relationship

$$\varphi_\Lambda(\mathbf{x}) = \varphi_0 - k\left[\sqrt{(x+H)^2 + y^2 + L^2} + \sqrt{(x-H)^2 + y^2 + L^2}\right] \quad (6.14)$$

where $\varphi_0 = const$. In the paraxial approximation

$$\varphi_\Lambda(\mathbf{x}) \cong \varphi_0 - \frac{x^2+y^2}{L} + \frac{H^2}{L} \qquad (6.15)$$

and we obtain the analog of a reflecting thin lens located on an optical axis and having focal length *L/2*.

The size *d* of the planar optical element M will be chosen in accordance with the size *2a* of the fiber's core using the equation

$$\exp\left[-\left(\frac{a}{\sigma_0}\right)^2\right] = \exp\left[-\left(\frac{d/2}{\sigma}\right)^2\right]$$

that describes the limitation of the beam taking place at the ends at the same level of the intensity fall, where $\sigma = \sigma_0\sqrt{1+l^2/z_0^2}$ is the radius of the Gaussian beam in the planar *M*. Thus we have the relation

$$d = 2a(\sigma/\sigma_0).$$

An essential parameter here is the feature size Δ of the Fresnel zones of the reflecting lens. Under the condition *H/l*<<1 for the lens specified by Eq.(6.14), we find that the minimum width of microrelief zone is

$$\Delta = \frac{\lambda}{2}\sqrt{1+\left(\frac{L}{d/2}\right)^2} \cong \frac{\lambda}{2NA},$$

where *NA* is the numerical aperture of the waveguide. The values Δ for various *l* and *d* are given in the Table 6.1. One can see that for the fibers, which are really employed and characterized by *NA*- 0.1-0.25, and for $\lambda \approx 1$ the value amounts to the order of several microns.

## 6.5. EXPERIMENTAL INVESTIGATION OF MICROSHIFT SENSOR EMPLOYING MODANS

The use of modans makes it possible to develop the amplitude fiberoptic sensors characterized by the enhanced and controlled sensitivity[Golub, 1991c]. In the sensor of microshifts depicted in Fig. 6.3 the improvement of sensitivity is expected to be due to the fact that the intensity of a single selectively excited mode in the optical fiber sensitive to the microbend is used as the information signal. This differs from conventional fiberoptic sensors based on measuring the total intensity of light beam. The optical fiber *F* in use with the length of 1200 mm, that has the parabolic profile of the refractive index, experiences a single quasi-sine microbend characterized by the varying amplitude *h* (Fig. 6.4). The microbend is produced by the working membrane 4 having one projection and by the reference membrane 6 having two projections.

**Table 6.1**
**The sensitivity $S_{pl}$ of mode-selective fiberoptic sensor.**

| *p, l* | (0,0) | (1,0) | (2,0) |
|---|---|---|---|
| $S_{pl}$, Theory | 1.0 | 1.9 | 2.3 |
| $S_{pl}$, Experiment | 1.0 | 1.8 | 2.1 |

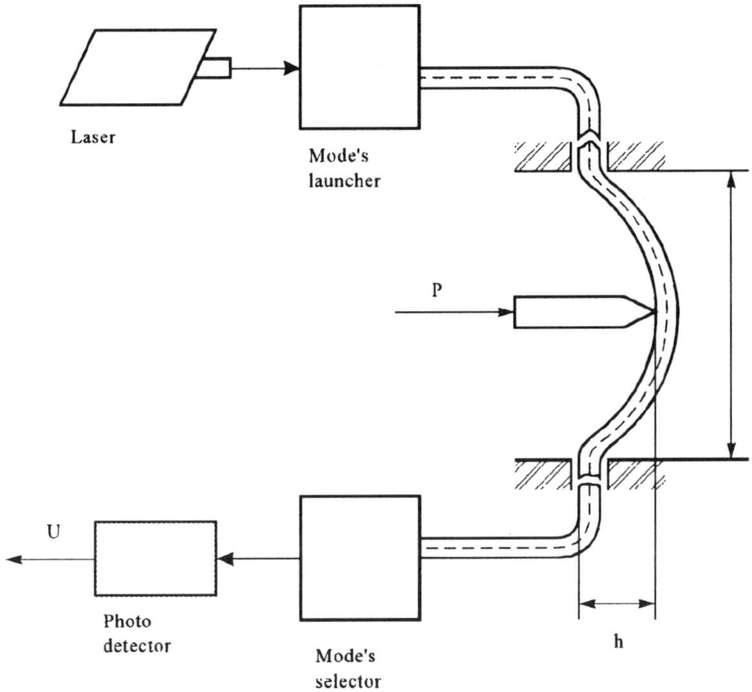


**Figure 6.3.** Block-diagram for the fiber sensor of microshifts. $P$ - mechanical microshifts, $h$ - amplitude of fiber's deflection, $U$ - output electrical voltage.

The distance 1.0 mm between projections has been chosen so that it should ensure the resonance mode [Unger, 1977]. Both membranes are fixed between the frame 3 and the lids 5 and 8.

The hollow screw 9 produces the initial deformation of a fiber. The measurements of the working membrane shifts are carried out with the help of the clockwork indicator 1. The fiber $F$ passes through two metallic micro-tubes 2 embossed into the frame 3 and filled in with the silicone hermetic filler 7.

The selective excitation of separate transverse modes in a fiber takes place in a mode formator that incorporates the laser $L$, the collimator $O_1$-$O_2$, the modan $M_1$ and objective $O_3$ The modan's $M_1$ complex transfer is matched to one of the fiber's modes.

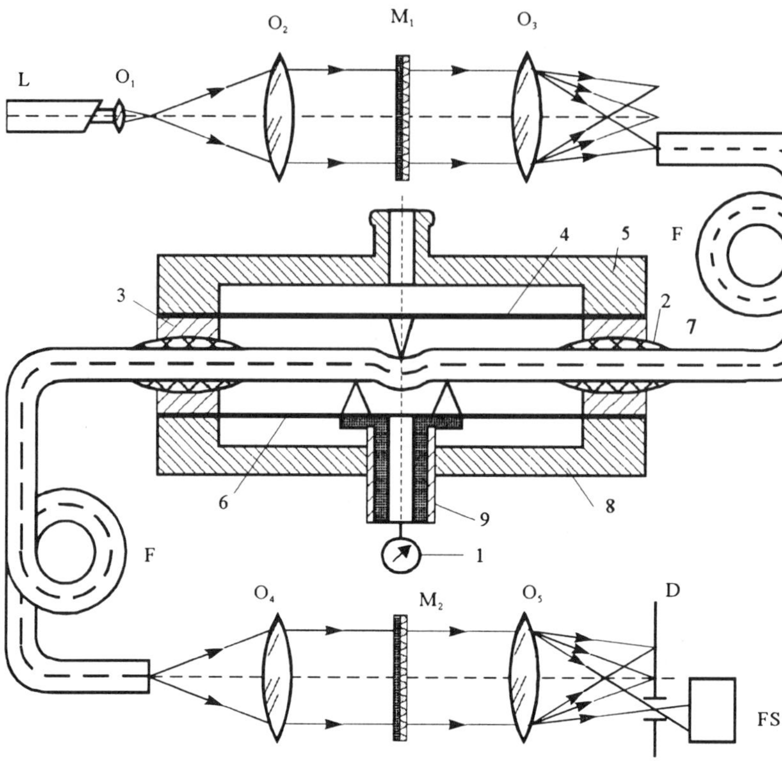


**Figure 6.4.** Optical scheme for the fiber optical sensor of microshifts. $L$ - laser; $O_1, O_2, O_5$ - objectives; $O_3, O_4$ - microobjectives; $M_1, M_2$ - multibeam modan; $F$ - fiber; $D$ - diaphragm; $FS$ - photoreceiver; *1* - clock-work indicator of membrane's shift; *2* - metallic microtubes; *3* - frame of the sensor; *4* - working membrane; *5* - upper lid; *8* - bottom lid; *6* - reference membrane; *7* - silicone hermetic filter; *9* - hollow screw for initial deformation of fiber. (Modified from Golub M.A., Sisakyan I.N., Soifer V.A., Uvarov G.V., Mode-selective fiber sensors operating with computer generated optical elements, *Proceedings SPIE, OFSC'91*, 1572, 101, 1991).

Selective separation of the definite mode at the fiber output and the registration of mode's intensity are conducted in the mode analyzer, the former consisting of the objective $O_4$, the second modan $M_2$, the $O_5$ objective and the photo-receiver $FS$ with the diaphragm $D$. The modan's $M_2$ complex transfer is matched to the mode configuration that occurs at the fiber input.

The selective excitation of the mode at the input together with the selective detection at the fiber's output ensure the singlemodal operation of the multimodal fiber. The number of mode to operate on is under our control by changing modan or modan's branch. So we can use several single modes with the same fiber.

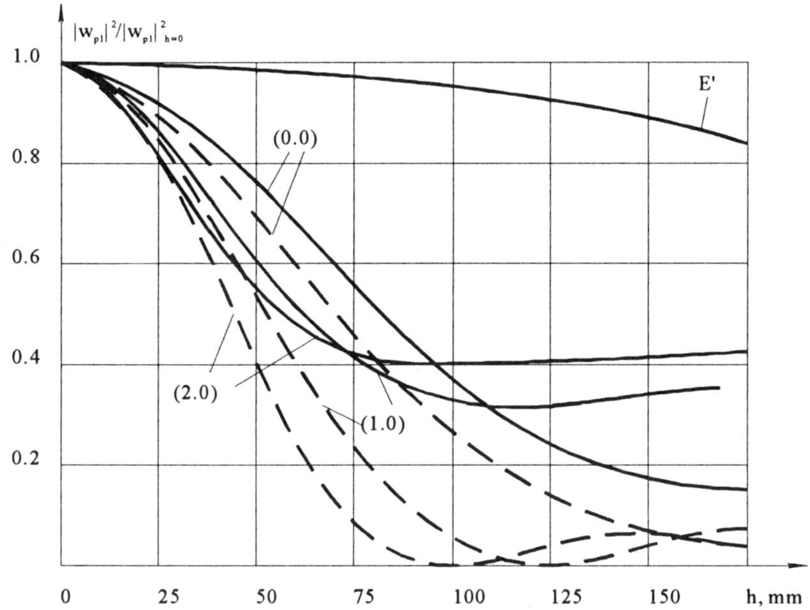


**Figure 6.5.** The sensor's output signal versus the shift $h$ of working diaphragm $E'$ - integral light flux emerging from the fiber; pairs of numbers (0,0), (1,0), (2,0) mark the curves corresponding to proper mode selection. (Modified from Golub M.A., Sisakyan I.N., Soifer V.A., Uvarov G.V., Mode-selective fiber sensors operating with computer generated optical elements, *Proceedings SPIE, OFSC'91*, 1572, 101, 1991).

The variations of the scale as well as the radial, axial, and angular adjustment (see section 5.6) ensure the complete matching of the modal configuration to the mode of the fiber. Fig. 6.5 shows the output signal (experimental and theoretical plots) proportional to the separate mode intensity versus the value of the fiber-bending deflection $h$. The output signal that is power of single mode is normalized to the maximum value upon the zero bending deflection. The variations of the radiation total flux E' for the Gaussian input beam is also shown in Fig. 6.5. As is seen from the experimental curves in Fig. 6.5, the variations of normalized modal powers measured with the help of modans, fall within the interval 0.4-1, whereas the variations of the total light flux E take by within the smaller interval of 0.85 -1, given the same values of the microbending deflections of 0-80 μm. There is a portion of the experimental curves in Fig. 6.5 that is nearly constant (beginning with $h$=80 μm), different from theoretical predictions (see [Grau, 1980]). It corresponds to the inoperative interval and results from the effects of misadjustment of the optical set-up, perturbations of modans.

Figure 6.6 illustrates the modal sensitivity

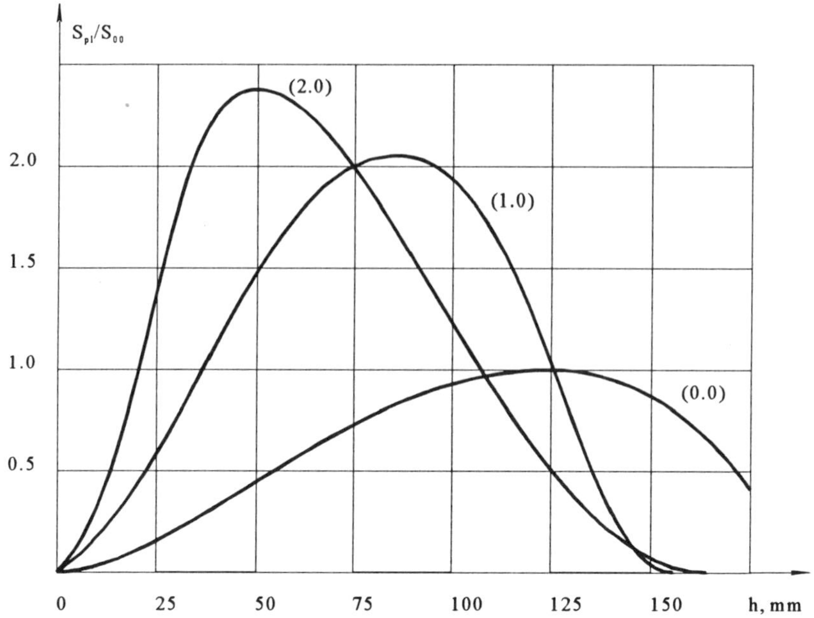


**Figure 6.6.** The sensor's sensitivity versus the shift *h* of working diaphragm. Pairs of numbers (0,0), (1,0), (2,0) mark the curves corresponding to proper mode selection. (Modified from Golub M.A., Sisakyan I.N., Soifer V.A., Uvarov G.V., Mode-selective fiber sensors operating with computer generated optical elements, *Proceedings SPIE, OFSC'91*, 1572, 101, 1991).

$$S_{pl} = \frac{\partial |w_{pl}|^2}{\partial h}$$

with respect to the microbending deflections *h* normalized to the maximum sensitivity of the principal mode (0,0). The analysis of the characteristics in Figs. 6.5 and 6.6 shows that the maximum sensitivity of modes in terms of the bending deflection occurs approximately at the level of the output signal within the range of 0.4 - 0.6. We failed reach such a level of attenuation experimentally since the destruction of a fiber took place if h becomes equal to 175 μm.

Let us estimate the ratio of mode-selective sensor sensitivity to the sensitivity of conventional fiber sensor measuring a total flux. The value 175 μm of the bending deflection will be chosen for the comparison. This causes the maximum output signal attenuation of 0.85 in the conventional sensor. The analysis of the plots in Fig. 6.6 and Table 6.1 shows the sensitivity of the principal mode (0,0) to be 3.9 times higher than the sensitivity of the total flux. The sensitivity of the (1,0) mode is 2.1 as high and of the (2,0) mode is 2.4 as high as that of the principle mode (see Table 6.1).

Thus the mode-selective fiberoptic sensor ensures 4-8 times increase in switchable sensitivity if compared with conventional fiberoptic sensor.

## Appendix 1.

## THE PROOF OF THE PROPERTIES OF ORTHOGONAL MODES

Let us consider in more detail the optimal properties of modes discussed of sections 1.1. We are going to make use of the designations in section 1.1 and 2.2.

**A. Conservation of orthogonality property**

The $\psi_{\mathbf{p}}$ modes are orthogonal, Eq.(2.5), in the transverse cross-section of a beam. While propagating, the modes conserve their orthogonality since the mode $\hat{P}\psi_{\mathbf{p}}$ just after propagation obey the equation

$$\left(\hat{P}\psi_{\mathbf{p}},\hat{P}\psi_{\mathbf{p}'}\right)=\left(\gamma_{\mathbf{p}}\psi_{\mathbf{p}},\gamma_{\mathbf{p}'}\psi_{\mathbf{p}'}\right)=\gamma_{\mathbf{p}}\gamma_{\mathbf{p}'}^{*}\left(\psi_{\mathbf{p}},\psi_{\mathbf{p}}\right)=\left|\gamma_{\mathbf{p}}\right|^{2}\delta_{\mathbf{pp}'}$$

It is worth noting that the modes are one of but indeed unique set of the 2-D base-functions conserving their orthogonality during propagation with respect to $\hat{P}$ operator. Suppose for the proof of such a statement, that there is some different orthogonal basis $\{e_{\mathbf{p}}(\mathbf{x})\}$ $\left(e_{\mathbf{p}},e_{\mathbf{p}'}\right)=\delta_{\mathbf{pp}}$ conserving the orthogonality in the process of propagation, that is, the following relationships are satisfied for all indices $\mathbf{p}'\neq\mathbf{p}$:

$$\left(\hat{P}e_{\mathbf{p}},\hat{P}e_{\mathbf{p}'}\right)=\left(\hat{P}^{*}\hat{P}e_{\mathbf{p}},e_{\mathbf{p}'}\right)=0$$

Owing to the completeness property of the basis $\{e_{\mathbf{p}}(\mathbf{x})\}$, one should conclude that

$$\hat{P}^{*}\hat{P}e_{\mathbf{p}}=\tilde{\lambda}_{\mathbf{p}}e_{\mathbf{p}}$$

that is, $\{e_{\mathbf{p}}\}$ are the eigenfunctions for the operator $Q=\hat{P}^{*}\hat{P}$. However, in accordance with Eqs.(1.5), (1.2) this means that $\tilde{\lambda}_{\mathbf{p}}=\lambda_{\mathbf{p}}$ and $e_{\mathbf{p}}=\psi_{\mathbf{p}}$ that proves the uniqueness property of modes. Thus, we have proved the first property of modes, listed in the section 1.1.

**B. Mean square errors of truncation of the modal expansions**

If we assume that, with respect to the waveguides under study, the propagation operator possesses the mathematical property of normality compactness (see [Ahiezer, 1966]), we find that a set of eigenfunctions $\psi_{\mathbf{p}}$ is a basis in $L_{2\rho_0}(G)$ space of functions. Thus every field can be uniquely expanded in terms of the $\psi_{\mathbf{p}}$ modes.

Let the field $w(\mathbf{x})$ be represented as an expansion in terms of modes

$$w(\mathbf{x})=\sum_{\mathbf{p}}w_{\mathbf{p}}\psi_{\mathbf{p}}(\mathbf{x}),\quad \mathbf{x}\in G \tag{Ap1.1}$$

where, by virtue of the modes' orthogonality, Eq.(2.5), the mode coefficients are specified by the integral

$$w_{\mathbf{p}} = \left(w, \psi_{\mathbf{p}}\right) = \int_G w(\mathbf{x}) \psi_{\mathbf{p}}^*(\mathbf{x}) \rho_0(\mathbf{x}) d^2\mathbf{x} \qquad \text{(Ap1.2)}$$

The set of the coefficients $\{w_{\mathbf{p}}\}$ can be treated as a result of the direct modal transform with respect to the function $w(\mathbf{x})$ and can be referred to as the spectrum of transverse modes. An inverse modal transformation sets up the correspondence between the mode spectrum $\{w_{\mathbf{p}}\}$ and the field $w(\mathbf{x})$, Eq.(Ap1.1).

On propagating by one pass in a waveguide medium, the "launching" beam, Eq.(Ap1.1), converges to the "output" beam $W = \hat{P}w$

$$W(\mathbf{x}) = \sum_{\mathbf{p}} W_{\mathbf{p}} \psi_{\mathbf{p}}(\mathbf{x}), \qquad \text{(Ap1.3)}$$

where

$$W_{\mathbf{p}} = \gamma_{\mathbf{p}} w_{\mathbf{p}} \qquad \text{(Ap1.4)}$$

Relationship (Ap1.4) makes it possible to call the series $\{\gamma_{\mathbf{p}}\}$ the transfer function of linear system $\hat{P}$, when using the mode basis $\{\psi_{\mathbf{p}}\}$. The transfer function after $m_0$ passes equals $\{\gamma_{\mathbf{p}}^{m_0}\}$.

The Parseval equality applied to the orthogonal expansions, Eqs.(Ap1.1) and (Ap1.3), gives us

$$\|w\|^2 = \sum_{\mathbf{p}} \left|w_{\mathbf{p}}\right|^2, \quad \|W\|^2 = \sum_{\mathbf{p}} \left|W_{\mathbf{p}}\right|^2 \qquad \text{(Ap1.5)}$$

Thus, the powers of the launching and output light beams

$$E = \|w\|^2 = \int_G |w(\mathbf{x})|^2 d^2\mathbf{x}$$

$$E\gamma = \|W\|^2 = \left(\hat{P}w, \hat{P}w\right) = \left(\hat{P}^*\hat{P}w, w\right) = \left(\hat{Q}w, w\right) \qquad \text{(Ap1.6)}$$

are separated into the powers of the independent modes $\left|w_{\mathbf{p}}\right|^2$ and $\left|W_{\mathbf{p}}\right|^2$ or into the powers of individual modes and satisfy, according to Eq.(Ap1.4), the following equation

$$\left|W_{\mathbf{p}}\right|^2 = \lambda_{\mathbf{p}} \left|w_{\mathbf{p}}\right|^2. \qquad \text{(Ap1.7)}$$

Eq (Ap1.7) allows the $\{\lambda_{\mathbf{p}}\}$ sequence to be called the energy transfer function.

The sequences

$$\left\{\frac{|w_{\mathbf{p}}|^2}{\|w\|^2}\right\}\ \left\{\frac{|W_{\mathbf{p}}|^2}{\|W\|^2}\right\} \tag{Apl.8}$$

describe the power distribution of light beams $w$ and $W$ between modes. Also, the beams $w(\mathbf{x})$ and $W(\mathbf{x})$ can be specified by a finite number of constituent modes $\psi_{\mathbf{p}}$, $\mathbf{p} \in K$, with the relative mean-square deviations

$$\delta_0^2 = 1 - \sum_{\mathbf{p} \in K} \frac{|w_{\mathbf{p}}|^2}{\|w\|^2} \tag{Apl.9}$$

and

$$\delta_\gamma^2 = 1 - \sum_{\mathbf{p} \in K} \frac{|W_{\mathbf{p}}|^2}{\|W\|^2} \tag{Apl.10}$$

**C. Minimum losses in power for modes during propagation**

In a passive waveguide with no energy supplied, the following inequality takes place

$$E_\gamma = \left(\hat{Q}w, w\right) \le (w, w) \equiv E$$

Relative diffraction and other losses during one full pass are equal to

$$\alpha = \frac{\hat{E}_\gamma - E}{E} = 1 - \frac{\left(\hat{Q}w, w\right)}{\|w\|^2} \tag{Apl.11}$$

Substituting Eq.(Apl.1) into Eq.(Apl.11) and making use of Eqs.(1.6) and (1.5) we obtain the convenient equation

$$\alpha = \sum_{\mathbf{p}} \alpha_{\mathbf{p}} \frac{|w_{\mathbf{p}}|^2}{\|w\|^2} \tag{Apl.12}$$

that relates the losses α of the beam $w$ to the losses $\alpha_{\mathbf{p}}$ of each its mode and to the power distribution between modes. When computing losses in $m_0$ passes of a beam in a waveguide, one should take $\alpha_{\mathbf{p}}^{m_0}$ instead of $\alpha_{\mathbf{p}}$ in Eq.(Apl.12). The modes of higher orders for which

$$\alpha_{\mathbf{p}} \cong 1, \quad |\gamma_{\mathbf{p}}| \cong 0$$

are practically unable to propagate in a waveguide medium. If the field $w$ containing modes of higher order will be launched into a waveguide, the modes will be attenuated; that is, the generalized spectrum $\{W_{\mathbf{p}}\}$ will, owing to Eqs.(Ap1.7) and (Ap1.4) be cut off by the power transfer function $\lambda_{\mathbf{p}}$. The light flux lost on the modes of higher order, Eq.(Ap1.10), makes a pronounced portion of the beam general losses, Eq.(Ap1.12).

Let us introduce number $\mathbf{p}_0$ of the mode of the lowest order included into the beam $w$, that is,

$$\lambda_{\mathbf{p}_0} = \max_{|w_{\mathbf{p}}| \neq 0} \lambda_{\mathbf{p}}$$

and

$$\alpha_{\mathbf{p}_0} \leq \alpha_{\mathbf{p}}$$

The value α, Eq.(Ap1.12), allows the following estimate to be done

$$\alpha \leq \alpha_{\mathbf{p}_0}, \tag{Ap1.13}$$

the former expressing, according to Eq.(Ap1.11), the well-known extremum property of eigenvalues of the operator $\hat{Q}$ (see [Korn, 1968]). Since in Eqs.(Ap1.11), (Ap1.12) and (Ap1.13) the beam $w$ is taken as arbitrary, with $\alpha_{\mathbf{p}_0}$ to be the losses of the mode $\psi_{\mathbf{p}_0}$, the inequality (Ap1.13) makes it possible to formulate a remarkable property (property 2, section 1.1) of modes as the light field special configurations closely connected to the waveguide medium.

Thus, in order to minimize losses, it seems worthy trying to concentrate the power of the launched beam on the modes of the lowest order characterized by the minimum losses.

**D. Optimal basis for finite-term approximation of orthogonal expansion.**

Here, we follow the teaching book of [Daugavet, 1977]. The property 3 mentioned in section 1.1 and dealing with the field approximation in a beam cross-section by the finite number of modes, follows from a mathematical theorem that is formulated below and represents the generalization of the analogous theorem for self-conjugated operator that has been proved in mathematical teaching books (see [Ahiezer, 1966]).

Let us consider the approximations of the $f$ function by a finite number of the base-functions $\{\psi_k\}$, where $k$ is the number of modes that are linearly ordered according to the increasing losses. For any given orthonormal system of the functions $\{e_k(\mathbf{x})\}$ specified in $L_{2,\rho_0}(G)$ functional space, the error of the approximation of the $f(\mathbf{x})$ function by the truncated orthogonal series

$$f(\mathbf{x}) = \sum_{k=1}^{\infty}(f, e_k)e_k(\mathbf{x}) \cong \sum_{k=1}^{L}(f, e_k)e_k(\mathbf{x})$$

can be found from the equation

$$E_L(f, \{e_k\}) = \left( \|f\|^2 - \sum_{k=1}^{L} |(f, e_k)|^2 \right)^2 \quad \text{(Ap1.14)}$$

For the definite class $F \subset L_{2,\rho_0}(G)$ of the $f$ functions, the error is determined from the "worst" approximation within the members of $F$:

$$\mathrm{E}_L(F, \{e_k\}) = \sup_{f \in F} E_L(f, \{e_k\}) \quad \text{(Ap1.15)}$$

If, in addition, we try to select the basis that is matched to the class of functions under approximation and able to provide the least approximation error, there arises a concept of "diameter":

$$d_L(F, \{\psi_k\}) = \inf_{\{e_k\}} \mathrm{E}_L(F, \{e_k\}), \quad \text{(Ap1.16)}$$

where $\{\psi_k\}$ is the optimum basis matched to the class $F$ of the functions under approximation.

We consider the operator $\hat{P}$ together with its eigenvalues $\gamma_k \equiv \gamma_{\mathbf{p}}$ arranged in order of decreasing $|\gamma_{\mathbf{p}}|$:

$$|\gamma_0| \geq |\gamma_1| \geq |\gamma_2| \geq \ldots \geq |\gamma_L| \geq |\gamma_{L+1}| \geq \ldots \quad \text{(Ap1.17)}$$

Thus the eigenvalue of $\gamma_{L+1}$ correspond to the mode, which is next to the $L$-th mode in terms of increasing losses.

Theorem (About the diameter of the ellipsoid in a Hilbert space) Let us assume that $\hat{P}$ is a compact normal linear operator in a separable Hilbert space. The set of functions

$$F = \left\{ f : f = \hat{P}g,\ \|g\|^2 \leq E \right\} \quad \text{(Ap1.18)}$$

represents an ellipsoid.

In that case for any $L$=1,2,.. we have the equality

$$d_L(F, \{\psi_k\}) = \sqrt{E}\,|\gamma_{L+1}|, \quad \text{(Ap1.19)}$$

where $\gamma_{L+1}$ the eigenvalue number ($L$+1) of the operator $\hat{P}$.

The first eigenfunctions $\{\psi_k\}_1^L$ of the operator $\hat{P}$ are the optimal base functions for the class $F$.

Proof: Let us first prove that

$$\mathrm{E}_L(F,\{\psi_k\}) \leq M|\gamma_{L+1}|, \quad M=\sqrt{E}, \tag{Ap1.20}$$

where

$$\hat{P}\psi_k = \gamma_k \psi_k, \quad k=1,2,... \tag{Ap1.21}$$

Owing to the normality of the $\hat{P}$ operator, one can choose the eigenfunctions $\{\psi_k\}_1^\infty$ orthonormal, the eigenvalues $\gamma_k$ being complex. Let us have that $f \in F$. For the function $g$ with $\|g\|^2 \leq E$ we can write orthogonal expansion as

$$g = \sum_{k=1}^{\infty} \gamma_k \psi_k, \quad g_k = (g, \psi_k) \tag{Ap1.22}$$

with the corresponding Parseval equality in the form

$$\|g\|^2 = \sum_{k=1}^{\infty} |g_k|^2$$

Applying the $\hat{P}$ operator to Eq.(Ap1.22) and allowing for Eq.(Ap1.21), we obtain the expansion

$$f = \hat{P}g = \sum_{k=1}^{\infty} \gamma_k g_k \psi_k$$

Making use of the expression that specifies the error in the approximation by the finite part of the orthogonal series, we find according to Eq.(Ap1.17) that

$$E_L(f,\{\psi_k\}) = \left(\sum_{k=L+1}^{\infty} |\gamma_k|^2 |g_k|^2\right)^{1/2} \leq |\gamma_{L+1}| \left(\sum_{k=L+1}^{\infty} |g_k|^2\right)^{1/2} \leq |\gamma_{L+1}| \|g\| \leq M|\gamma_{L+1}| \tag{Ap1.23}$$

By virtue of the arbitrariness of $f$ in Eq.(Ap1.23) and taking sup, operation we come directly to Eq.(Ap1.20). Furthermore, from the definition of the diameter, Eq.(1.16) we have

$$d_L(F,\{\psi_k\}) \leq \mathrm{E}_L(F,\{\psi_k\}) \leq M|\gamma_{L+1}| \tag{Ap1.24}$$

Let us prove now the opposite inequality

$$d_L(F,\{\psi_k\}) \geq M|\gamma_{L+1}| \tag{Ap1.25}$$

Let $\{e_k\}_1^L$ be an arbitrary orthonormal basis. For the following system of the equations written with respect to $\alpha_j$, $j=\overline{1,L}$ variables

$$\sum_{j=1}^{L+1} \gamma_j \alpha_j (\psi_j, e_k) = 0, \quad k = \overline{1,L}, \tag{Ap1.26}$$

the number of the unknown variables is more by one than the number of the equations. thus if we put, for example, $\alpha_{L+1} = 1$, than we can find the remaining numbers $\alpha_1,...,\alpha_L$ from Eq.(Ap1.26) and obtain a non-zero solution $(\alpha'_1,...,\alpha'_L,\alpha'_{L+1})$. If we perform the normalization

$$\alpha_j = M \frac{\alpha'_j}{\left( \sum_{j=1}^{L+1} |\alpha_j|^2 \right)^{1/2}}$$

we obtain the solution of Eq.(Ap1.26) for which

$$\sum_{j=1}^{L+1} |\alpha_j|^2 = M^2 \tag{Ap1.27}$$

Let us assume $g = \sum_{j=1}^{L+1} \alpha_j \psi_j$. It follows from Eq.(Ap1.27) that

$$\|g\| = \left( \sum_{j=1}^{L+1} |\alpha_j|^2 \right)^{1/2} = M.$$

Therefore, the function $f = \hat{P}g$ has the following form

$$f = \hat{P}g = \sum_{j=1}^{L+1} \gamma_j \alpha_j \psi_j \in F \tag{Ap1.28}$$

and according to Eqs.(Ap1.26) and (Ap1.28) we find that

$$(f, e_k) = 0, \quad k = \overline{1,L}$$

At this stage, for the orthogonal approximation

$$f \cong \sum_{k=1}^{L} (f, e_k) e_k$$

we can find the error

$$E_L(f,\{e_k\}) = \left( \|f\|^2 - \sum_{k=1}^{L} |(f,e_k)|^2 \right)^{1/2} = \|f\| =$$
$$= \left( \sum_{j=1}^{L+1} |\gamma_j \alpha_j|^2 \right)^{1/2} \geq |\gamma_{L+1}| \left( \sum_{j=1}^{L+1} |\alpha_j|^2 \right)^{1/2} = M |\gamma_{L+1}| \qquad \text{(Ap1.29)}$$

Taking a sup-operation with respect to both sides of Eq.(Ap1.29), we obtain

$$\mathrm{E}_L(f,\{e_k\}) \geq M|\gamma_{L+1}| \qquad \text{(Ap1.30)}$$

Since the basis $\{e_k\}$ is arbitrary, we can take inf-operation relative to both sides of Eq.(Ap1.30) and obtain inequality (Ap1.25). From Eqs.(Ap1.25) and (Ap1.24) it follows that Eq.(Ap1.19) holds true. This was to be shown.

Thus, when at the waveguide input a beam $w$ with the power no more than $E$ is launched, the best approximation of the output beam $\hat{P}w$ is by first $L$ modes $\{\psi_p\}$ of the waveguide that are the eigenfunctions of the propagation operator $\hat{P}$. The best approximation can be physically interpreted as the minimization of the power of remaining modes of higher order not used in the representation of the beam $\hat{P}w$. According to Eq.(Ap1.19) the power of the remaining modes equals:

$$\Delta E = E\lambda_{L+1} \quad \left(\lambda_{L+1} = |\gamma_{L+1}|^2\right). \qquad \text{(Ap1.31)}$$

# Appendix 2.

## SINUSOIDAL MODES IN A REAL FORM

The basefunctions corresponding to sinusoidal modes are discribed by the following equations:

$$\psi_{\mathbf{p}}(\mathbf{x}) \equiv \psi_{pl}(x,y) = \psi_p(x)\tilde{\psi}_l(y) \quad \text{(Ap2.1)}$$

$$\psi_p(x) = E_{xp}\sin[z_p(\tau_1)\frac{x}{a} + \alpha_p] \quad \text{(Ap2.2)}$$

$$\tilde{\psi}_l(y) = E_{yl}\sin[z_l(\tau_2)\frac{y}{b} + \alpha_l], \quad \text{(Ap2.3)}$$

where p,l=0,1,2,... . Here $E_{xp}, E_{yl}$ are the normalization constants.

The matrix elements of perturbations for the functions in Eqs.(Ap2.1) through (Ap2.3) can be represented in the form

$$H_{plp'l'}^{(pl)} = -\delta_{plp'l'} + \int_{-a}^{a}\int_{-b}^{b} \varphi_{pl}(x,y)\Psi_{p'l'}(x,y)\,dxdy \quad \text{(Ap2.4)}$$

Since the separation of the variables x and y takes place in the base-functions Eq (Ap2.1) and in the function $\chi(x,y)$ in Eq.(4.38), the perturbed function $\varphi_{pl}$ also has separable variables

$$\varphi_{pl}(x,y) = \varphi_p(x)\varphi_l(y),$$

where

$$\varphi_p(x) = \sum_{j=1}^{N}\varphi_p(x_j)\,rect\left(\frac{x - x_j}{\delta}\right). \quad \text{(Ap2.5)}$$

From Eqs.(Ap2.5) and (Ap2.2) we find that

$$\int_{-a}^{a}\varphi_p(x)\psi_{p'}(x)dx = E_{xp}E_{xp'}\sum_{j=1}^{N}\int_{x_{j-1}}^{x_j}\sin\left(\frac{z_p x_j}{a} + \alpha_p\right)\cdot\sin\left(\frac{z_{p'}x}{a} + \alpha_{p'}\right)dx =$$

$$= \frac{aE_{xp}E_{xp'}}{z_{p'}}\sin\left(\frac{z_{p'}}{N}\right)\cdot$$

$$\cdot\sum_{j=1}^{N}\left[\cos\left(\frac{z_p x_j}{a} - \frac{z_{p'}}{a}\frac{x_j + x_{j-1}}{2} + \alpha_p - \alpha_{p'}\right) - \cos\left(\frac{z_p x_j}{a} + \frac{z_{p'}}{a}\frac{x_j + x_{j-1}}{2} + \alpha_p - \alpha_{p'}\right)\right]$$

(Ap2.6)

Making use of the well-known trigonometric relationship of summation

$$\sum_{n=1}^{N}\cos(\alpha+n\beta)=\cos\left(\alpha+\frac{\beta}{2}+\frac{N\beta}{2}\right)\sin\left(\frac{N\beta}{2}\right)\bigg/\sin\left(\frac{\beta}{2}\right), \qquad \text{(Ap2.7)}$$

carrying out transformations, substituting Eq.(Ap2.6) into Eq.(2.4), and introducing the designations of Eqs.(Ap4.28) and (Ap4.29), we obtain the resulting Eq.(Ap4.27).

## Appendix 3.

## ANALYTICAL EQUATIONS FOR GAUSSIAN MODES EXCITATION IN OPTICAL FIBERS AND CAVITIES

Gauss-Laguerre or Gauss-Hermite modes are obtained in an optical fiber with the quadratic profile or the refractive index:

$$n^2(r) = n_1^2\left[1 - 2\Delta\left(\frac{r}{a}\right)^2\right], \quad r = \sqrt{x^2 + y^2} \le a, \quad \Delta = const$$

The radius $\sigma_0$ of the principal mode (with respect to a exp(−2) level of the intensity fall) can be determined by the well-known equations:

$$\sigma_0 = \frac{2a^2}{V}, \quad V = n_1 ka\sqrt{2\Delta}; \quad \beta_0 = -\frac{\sqrt{2\Delta}}{a}$$

The beam modal composition in a fiber results from the conditions of the fiber excitation and fiber bending. [Grau, 1980; de Fornel, 1983; Krivoshlykov, 1983] [Krivoshlykov, 1987] have developed analytical equations for modal coefficients.

For example, let the exciting Gaussian beam numbered $(p_1, l_1)$ have the parameter $\sigma_1$ instead of $\sigma_0$, the center displaced at $\Delta x, \Delta y$ relative to the fiber center, and the axis tilted in the plane $x,z$ at an angle θ with respect to the fiber axis. As [Krivoshlykov, 1983] [Krivoshlykov, 1987] have shown, the coefficients of expansion in terms of the Gaussian-Hermite modes are described in this case by the equations

$$w_{pl} = w_p^{(p_1)}\,\tilde{w}_l^{(l_1)},$$

where

$$w_p^{(p_1)} = w_0^{(0)} \cdot (p!\,p_1!)^{-1/2}\,\mathrm{H}_{pp_1}(\sigma, \tau)$$

$$w_0^0 = \chi \exp\left[-\left(\frac{\Delta x \cos\theta}{\sigma^2}\right)^2 + i\beta_0 \Delta x \sin\theta \frac{\sigma_1^2}{\sigma_2^2} - \frac{k}{16}\beta_0^2 \chi^2 \sin^2\theta\right]$$

$$\sigma = -\chi^2 \cos\theta\left(\frac{\Delta x}{\sigma_0} - i\frac{1}{2}\beta_0 \sigma_1 \sin\theta\right)$$

$$\tau = \frac{1}{2}\chi^2 \cos\theta\left(\frac{\Delta x \cos\theta}{\sigma_1} + \frac{i\beta_0 \sigma_1 tg\theta}{2}\right)$$

$$\sigma_2^2=\sigma_1^2+\sigma_0^2\cos^2\theta,\quad \chi^2=\frac{2\sigma_0\sigma_1}{\sigma_2^2}$$

where $H_{pp1}(\sigma,\tau)$ is the Hermite polynomial [Erdelyi, 1953]

The values of $\tilde{w}_l^{(l_1)}$ can be found with the help of the similar equations at $\theta=0$ and with $\Delta y$ substituted instead of $\Delta x$.

Equations (1.3) through (1.16) are simplified upon the Gaussian-beam excitation and with $l_1=p_1=0$:

$$w_p^0=w_0^0(p!)^{-1/2}\left(\frac{\sigma_0^2\cos^2\theta-\sigma_1^2}{2\left(\sigma_1^2+\sigma_0^2\cos^2\theta\right)}\right)^{p/2}\cdot H_p\left\{\frac{\sqrt{2}\sigma_0\sigma_1\cos\theta}{\sqrt{\sigma_0^2\cos^2\theta-\sigma_1^2}}\left(\frac{\Delta x}{\sigma_1}\cos\theta+i\frac{\beta_0\sigma_1 tg\theta}{2}\right)\right\}$$

and $\tilde{w}_l^0$ can be expressed by a similar equation, but with $\theta=0$ and $\Delta y$ substituted instead of $\Delta x$. Here $H_p$ is the 1-D Hermite polynomial [Erdelyi, 1953]. [Krivoshlykov, 1987] have developed equations for the Gauss-Laguerre modes for the exciting beam numbered $p_1, l_1$ with parameter $\sigma_1$, and with the axial displacement $\mathbf{\Delta x}=(\Delta x,\Delta y)$ of

$$\Delta x=\Delta r\cos\varphi_0;\qquad \Delta y=\Delta r\sin\varphi_0$$

and with no tilt $(\theta=0)$:

$$w_{pl}=\delta_{ll_1}\left(\frac{2\sigma_0\sigma_1}{\sigma_0^2+\sigma_1^2}\right)^{l+1}\left(\frac{\sigma_0^2-\sigma_1^2}{\sigma_0^2+\sigma_1^2}\right)^{p-p_1}\left[p_1!(p+l)!\right]^{1/2}\left[(p_1+l)!\,p!\right]^{-1/2}P_{p_1}^{(p-p_1,l)}\left(1-2\left(\frac{\sigma_0^2-\sigma_1^2}{\sigma_0^2+\sigma_1^2}\right)^2\right)$$

where $P_{p_1}^{(p-p_1,l)}$ is the Jacobi polynomial [Erdelyi, 1953].

[Grau, 1980] have discussed the case of $\theta\neq0$ upon the Gaussian exciting beam $(l_1=0,p_1=0)$. The modal power distribution in the case of Gauss-Laguerre modes takes the form:

$$\left|w_{pl}\right|^2=\frac{1}{1+\delta_{0l}}\cdot\frac{p!}{(p+l)!}\cdot\frac{4\sigma}{(1+\sigma)^2}\cdot\exp\left(-\frac{\rho^2+\sigma S^2}{1+\sigma}\right)\cdot\left(\frac{\sigma}{1+\sigma}\right)^{2l}\cdot\left(\frac{1-\sigma}{1+\sigma}\right)^{2p}\cdot$$

$$\cdot\left[\left(\frac{\rho^2}{\sigma^2}+S^2+2\frac{\rho}{\sigma}S\sin\varphi_0\right)^l+\left(\frac{\rho^2}{\sigma^2}+S^2-2\frac{\rho}{\sigma}S\sin\varphi_0\right)^l\right]\cdot$$

$$\cdot\left|L_p^l\left[\frac{\sigma^2}{1-\sigma^2}\left(\frac{\rho^2}{\sigma^2}+S^2+2i\frac{\rho}{\sigma}S\cos\varphi_0\right)\right]\right|^2,$$

where $\sin\theta_{max}$ is the numerical aperture of the fiber under consideration

$$\sigma=\left(\frac{\sigma_1}{\sigma_0}\right)^2,\quad \rho^2=2\left(\frac{\Delta r}{\sigma_0}\right)^2,\quad S^2=V\left(\frac{\sin\theta}{\sin\theta_{max}}\right)$$

$L_p^l$ is the Laguerre polynomial.

The modal power distribution for the beams in fibers and cavities are analyzed in Appendix 3.

The laser light modal composition depends on a variety of parameters of the active medium and cavity's. The adjustment is crucial for this. The modal power distribution in the passive cavity depends on a number of factors including the diffractive losses. [Solimeno, 1986] cited useful equation of Slepian

$$\alpha_{pl} = \frac{2\pi(8\pi N)^{2p+l+1} e^{-4\pi N}}{p!(p+l)!}, \beta_{pl} = (2p+l+1)\pi/2$$

for the losses and phase shifts in confocal cavities with annular mirrors and large Fresnel numbers. Here 2a is the mirror diameter and l is the distance between the mirrors.

It has been shown by [Vainshtein, 1966] that the losses in a plane-mirror cavity with annular mirrors are described by the other equation:

$$\alpha_{pl} = \frac{8k_{lp}\delta(m+\delta)}{\left[(m+\delta)^2+\delta^2\right]^2}, \beta_{pl} = \frac{m}{4\delta}\alpha_{pl}$$

where $\delta = 0.824$, $k_{lp}$ is the p-th zero of $J_l(\cdot)$ Bessel function, and $m = \sqrt{8\pi N}$.

# Appendix 4.

# RANDOM-FIELDS EXPANSIONS WITH DECORRELATED COEFFICIENTS

## AP4.1 EXPANSION IN TERMS OF EIGENFUNCTIONS OF THE CORRELATIVE OPERATOR

Random optical fields arise when the laser radiation propagates through a perturbing medium. This medium can be, for instance, spatial ensemble of aerosol particles. Random field occurs also in the situation when the laser light is reflected from the irregular surface or propagates in an optical fiber with a nonuniform refractive index. Expansion of the deterministic optical field in terms of independently propagating modes allows us to reduce its investigation to the analysis of the coefficients of the expansion. With the aim of the convenient representation of random optical fields, the expansion coefficients should also be independent (or at least noncorrelated) in a statistical sense. Random-field expansions with decorrelated coefficients are sometimes called canonical expansions. It is known (see [Karhunen, 1947]) that there is a unique orthogonal basis that provides decorrelated expansion coefficients for the case of the random field with a given correlation function. This "canonical" basis, or the Karhunen-Loeve basis (see [Klovskiy, 1976]), consists of eigenfunctions of the integral equation with the kernel that represents the correlation function of the discussed field. Thus, with respect to the random optical fields, the eigenfunctions of a correlation operator are a matter of interest, whereas for deterministic optical fields we examine modes, which are the eigenfunctions of the propagation operator.

Here the Karhunen-Loeve basis is dealt with in terms of modal expansions and mode selection taking place in the course of measuring the expansion coefficients, and the results of computational and optical experiments are discussed. Also, we have investigated the perturbations of eigenfunctions that occur when these are realized in the form of computer-generated optical spatial filters (see [Golub, 1980]).

When using the Karhunen-Loeve expansion in application to the digital image processing, [Watanabe, 1967] has employed statistical estimates of a correlation function. When dealing with physical objects, it appears expedient to use the calculated correlation functions of the complex amplitude.

Let the complex amplitude $w(\mathbf{x}), \mathbf{x} \in G$ of an optical field in the light beam cross-section be random and characterized by the preset correlation function

$$R(\mathbf{x},\mathbf{x}') = < [w(\mathbf{x}) - \overline{w}(\mathbf{x})][w(\mathbf{x}) - \overline{w}(\mathbf{x}')]^{*} >$$

where $\overline{w}(\mathbf{x}) = < w(\mathbf{x}) >$.

$<\cdot>$ denotes the ensemble average for the realizations w(**x**)

Let us introduce the correlation operator $R$

$$\hat{R}\Psi(\mathbf{x}) = \int_G R(\mathbf{x},\mathbf{x}')\Psi(\mathbf{x}')d^2\mathbf{x}', \qquad \text{(Ap4.1)}$$

which is selfconjugate and positively defined (see [Ahiezer, 1966]). Let us represent the field $w(\mathbf{x})$ in the form of the expansion

$$w(\mathbf{x}) = \sum_{\mathbf{p}} w_{\mathbf{p}}\Psi_{\mathbf{p}}(\mathbf{x}) + w(\mathbf{x}), \quad < w_{\mathbf{p}} >= 0 \qquad \text{(Ap4.2)}$$

in terms of the complete system $\{\Psi_{\mathbf{p}}\}$of orthonormal functions of two variables (the weight function $\rho_0 \equiv 1$)

$$\int_G \Psi_{\mathbf{p}}(\mathbf{x})\Psi_{\mathbf{p}'}^*(\mathbf{x})d^2\mathbf{x} = (\Psi_{\mathbf{p}}, \Psi_{\mathbf{p}'}) = \delta_{\mathbf{pp}'} \qquad \text{(Ap4.3)}$$

The coefficients of the orthogonal expansion can be found from the equation

$$w_{\mathbf{P}} = (w, \Psi_{\mathbf{p}}) = \int_G w(\mathbf{x})\Psi_{\mathbf{p}}^*(\mathbf{x})d^2\mathbf{x} \qquad \text{(Ap4.4)}$$

Also, we can represent the field as a truncated series comprising the limit number $L$ of terms with the indices $\mathbf{p} \in K,\ |K| = L$

$$w(\mathbf{x}) \cong \hat{w}_L(\mathbf{x}) \equiv \sum_{\mathbf{p}\in K} w_{\mathbf{p}}\Psi_{\mathbf{p}}(\mathbf{x}) + \overline{w}(\mathbf{x}) \qquad \text{(Ap4.5)}$$

The remainder of the orthogonal series

$$w(\mathbf{x}) - \hat{\mathbf{w}}_L(\mathbf{x}) = \sum_{\mathbf{p}\in K} w_{\mathbf{p}}\Psi_{\mathbf{p}}(\mathbf{x})$$

has the power

$$\varepsilon_0^2(L) =< \int_G |w(\mathbf{x}) - \hat{w}_L(\mathbf{x})|^2 d^2\mathbf{x} >, \qquad \text{(Ap4.6)}$$

Owing to Eq.(Ap4.3), we can write down Eq.(Ap4.6) as follows

$$\varepsilon_0^2(L) = E - \sum_{\mathbf{p}\in K} \left|(\hat{R}\Psi_{\mathbf{p}}, \Psi_{\mathbf{p}})\right|, \qquad \text{(Ap4.7)}$$

where

$$E =< \int_G |w(\mathbf{x})|^2 d^2\mathbf{x} >= \int_G R(\mathbf{x},\mathbf{x})d^2\mathbf{x} \qquad \text{(Ap4.8)}$$

is the average power of the random field $w$,

$$<\left|w_{\mathbf{p}}\right|^2>=(\hat{R}\Psi_{\mathbf{p}},\Psi_{\mathbf{p}}) \qquad \text{(Ap4.9)}$$

is the average power of the **p**-th orthogonal component. As is seen from Eq.(Ap4.7), the error $\varepsilon_0^2(L)$ is minimized for a preset basis if the functions $\{\Psi_{\mathbf{p}}\}_1^\infty$ are arranged in such order that the functions with larger dispersions are located in the first places, Eq.(Ap 4.9)

As noted in section 1.1, conventional modes provide the minimum power of the approximation error of the complex amplitude by the truncated series as compared with other sets of orthogonal functions of two variables (see Appendix 1). With respect to the random fields, [Karhunen, 1947] has also posed the problem of choosing an orthogonal set of the functions $\{\Psi_{\mathbf{p}}\}$ capable of providing the minimum approximation error $\varepsilon_0^2$, Eq.(Ap4.7). The minimization of $\varepsilon_0^2$ is equivalent to the maximization of the quadratic forms $\left|(\hat{R}\Psi_{\mathbf{p}},\Psi_{\mathbf{p}})\right|$.

Let us begin with the maximization of $\left|(\hat{R}\Psi_0,\Psi_0)\right|$. Owing to the well-known property of eigenfunctions (see [Korn, 1968]), the maximum $\lambda_0$ of the value $\left|(\hat{R}\Psi_0,\Psi_0)\right|$ for $\|\Psi_0\|=1$ is achieved on the R operator's eigenfunction, which has the maximum eigenvalue $\lambda_0$.

Since Eq.(Ap4.7) implies the limitation in the form of orthogonality, we should minimize the next quadratic form $(\hat{R}\Psi_{\mathbf{p}},\Psi_{\mathbf{p}})$ allowing for such a limitation as $(\Psi_{\mathbf{p}},\Psi_0)=0$. Therefore, based on the extremum properties of eigenfunctions, we can conclude that $\mathbf{\Psi_p}$ is the next eigenfunction and so on. Thus, the eigenfunctions of the correlative operator that satisfy the equation (see [Karhunen, 1947])

$$\hat{R}\Psi_{\mathbf{p}}=\lambda_{\mathbf{p}}\Psi_{\mathbf{p}} \qquad \text{(Ap4.10)}$$

are seen to provide the best approximation of a random optical field by the truncated orthogonal series.

The orthogonal set of eigenfunctions, Eq.(Ap 4.1), for the correlation operator is referred to as the Karhunen-Loeve basis (see [Fu, 1968] and [Watanabe, 1967]) and the series in Eq.(Ap4.2) as the Karhunen-Loeve expansion. Note that each eigenvalue appears as many times as its multiplicity $r_0$ in a sequence, which is arranged in the decreasing order. With respect to the corresponding sequence of the base-functions, to the eigenvalue that reoccurs $r_0$-fold there corresponds. The set of $r_0$ eigenfunctions of R operator forming a basis in the $r_0$-dimensional eigen-subspace (see [Ahiezer, 1977]) corresponds to $r_0$-multiple eigenvalue. Furthermore, since there takes place the relationship

$$< w_{\mathbf{p}} w_{\mathbf{p}'}^* >= (\hat{R}\Psi_{\mathbf{p}}, \Psi_{\mathbf{p}'}) = (\lambda_{\mathbf{p}}\Psi_{\mathbf{p}}, \Psi_{\mathbf{p}'}) = \lambda_{\mathbf{p}}\delta_{\mathbf{pp}'}, \qquad \text{(Ap4.11)}$$

the coefficients of the expansion in Eq.(Ap4.2) in terms of eigenfunctions, Eq.(Ap4.10), are decorrelated (see 249), or even independent if we are dealing with the field characterized by the normal distribution function.

In particular, for $\mathbf{p} = \mathbf{p}'$, we derive from Eq.(Ap4.11) that

$$< |w_{\mathbf{p}}|^2 >= (\hat{R}\Psi_{\mathbf{p}}, \Psi_{\mathbf{p}}) = \lambda_{\mathbf{p}} \qquad \text{(Ap4.12)}$$

This means that the error (Ap4.7) can be written in the form

$$\varepsilon_0^2(L) = E - \sum_{\mathbf{p}\in K} \lambda_{\mathbf{p}} \qquad \text{(Ap4.13)}$$

Example. Let the correlation function be of biexponential-cosine, type $\mathbf{x} = (x, y)$, $\mathbf{x}' = (x', y')$, $|x| \le a$, $|x'| \le a$, $|y| \le b$, $|y'| \le b$,

$$R(\mathbf{x}, \mathbf{x}') = \sigma^2 \exp\left[-\frac{\tau_1}{a}(x - x') - \frac{\tau_2}{b}(y - y')\right] \cdot$$

$$\cdot \cos\left[\frac{\beta_1}{a}(x - x')\right] \cdot \cos\left[\frac{\beta_2}{b}(y - y')\right] \qquad \text{(Ap4.14)}$$

For $\beta_1 = \beta_2 = 0$, the base-functions of the Karhunen-Loeve expansion, see Fig. Ap4.1, have the form of Eqs.(Ap2.1) through Eq.(Ap2.3) and are characterized by the parameters

$$E_{xp} = \{a[1 + e_p(\tau_1)]\}^{-1/2}, \quad E_{yl} = \{b[1 + e_l(\tau_2)]\}^{-1/2}$$

$$e_p(\tau) = \frac{\tau}{\tau^2 + z_p^2(\tau)}, \quad \alpha_p = \frac{\pi}{2} \frac{1 + (-1)^p}{2},$$

where $z_p(\tau)$ is the root of the transcendental equation

$$z_p = \frac{\pi}{2}(p+1) - arctg\frac{z_p}{\tau}$$

at the interval

$$\frac{\pi}{2} p < z_p < \frac{\pi}{2}(p+1)$$

Let us indicate the first ten pairs of the indices (p,l) at $\tau_1 = \tau_2 = 2{,}3$ and $\beta_1 = \beta_2 = 0$: (0,0),(0,1),(1,0),(1,1),(0,2),(2,0),(1,2),(2,1),(0,3),(3,0).

The appearance of the mean-square deviation $\varepsilon_0^2$, Eq.(Ap4.7), for the continuous Karhunen-Loeve basis is shown in Fig. Ap4.2. For the aim of comparison, the similar curves for the "continuous" Walsh- and Fourier-bases are also presented. The curve for the Karhunen-Loeve basis is seen to lie below those for the Walsh- and Fourier-bases; that proves the property of the minimum error. The curves substantiate the fact that the error $\varepsilon_0^2$ $(L)$ can be made as small as needed by choosing large $L$.

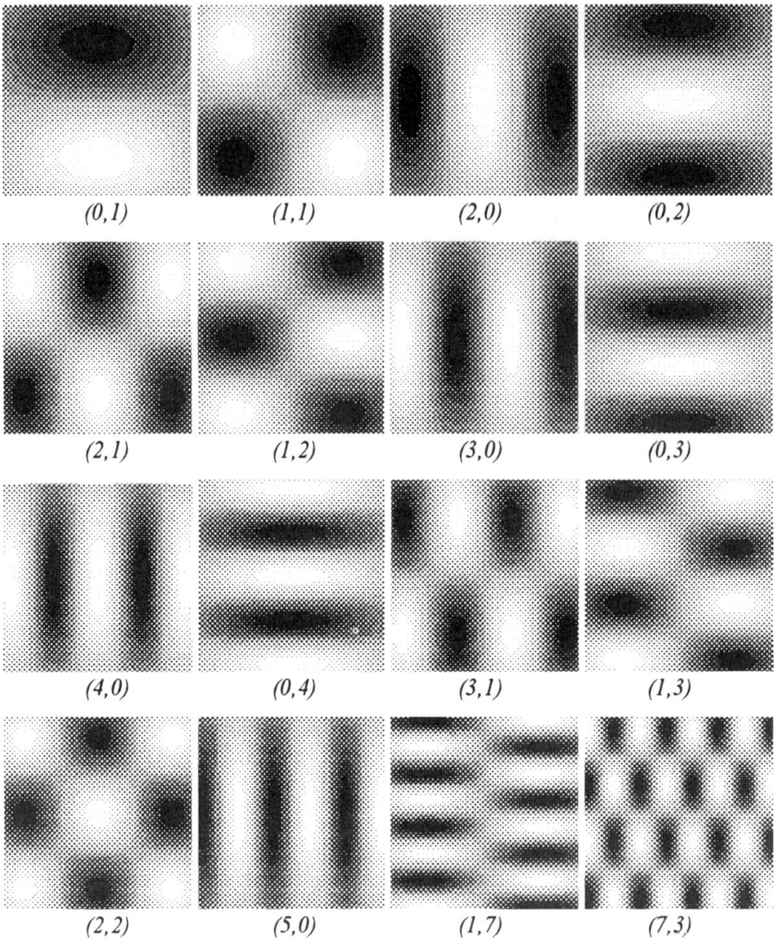


**Figure Ap.4.1.** Photographs of the spatial filters realizing the Karhunen-Loeve base-functions characterized by the biexponential correlation ($\tau_1 = \tau_2 = 2.3$).

The investigation of the expansions specified by Eqs.(Ap4.2) and (Ap4.10) and characterized by the decorrelated coefficients turns out to be convenient by the reason of both the independent character of the expansion terms and the most adequate representation of the field by the first terms of the expansion.

D.D.Klovskiy and V.A.Soifer [Klovskiy, 1976] have shown that application of the expansion with the decorrelated coefficients for the spatial transfer characteristic of medium makes it possible to reduce the problem of propagation in a random medium to the set of problems of light propagation in deterministic media.

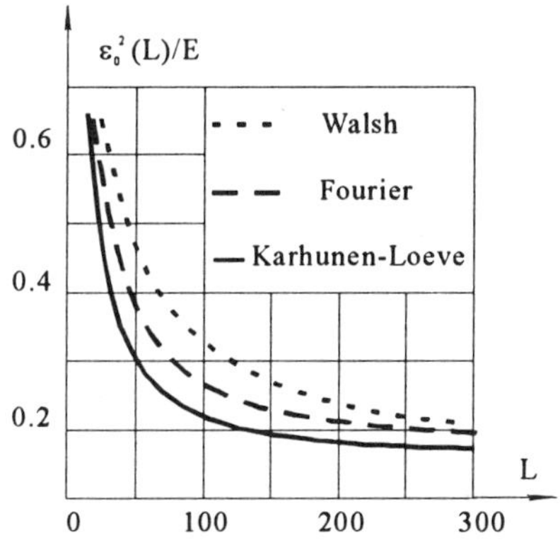


**Figure Ap.4.2.** Mean-square error in the random field representation through the truncated series of eigenfunctions. (Modified from Golub M.A., Soifer V.A., The stability of Karhunen-Loeve expansion and computer generation of optimal spatial filters, in *Spectral methods of information processing in scientific research*, Center for biological research, Pushino, 1980, 108 (in Russian)).

Another application of the expansions (Ap4.2) and (Ap4.10) is the proper registration of the modal composition of light beam under the random perturbations of the medium.

If a singlemodal Gaussian light beam is reflected from an irregular surface, there occurs its transformation into a multimodal beam with random modal coefficients. Building up the orthogonal functions from the correlative function of the beam and developing them in terms of the modes, we obtain those modal groups that have the maximum average power. Whereas prior to constructing the functions of Eq.(Ap4.10) it was difficult to separate the numbers of the modes being the most substantial in terms of their power.

## AP4.2 PERTURBATIONS OF THE KARHUNEN-LOEVE EXPANSION

Perturbations of the Karhunen-Loeve expansion are due to the sampling of its base functions taking place in accordance with the general conclusions of sections 4.1 and 4.2. In this case the eigenfunctions $\psi_p$ are affected by the perturbations $h_p$ and transformed into the "perturbed" base functions Eq.(Ap4.10). For the purposes of this section, let us consider the illuminating beam $w_0$ to be plane ($w_0$ = const) and the carrier frequency $\boldsymbol{\nu}_n^{(k)}$ equal to zero.

### A. Perturbation of base-functions

Under the perturbations of the base functions in the Karhunen-Loeve expansion we obtain the "perturbed" and weakly correlated coefficients

$$\hat{w}_{\mathbf{p}} = \int_G w(\mathbf{x})\varphi^*_{\mathbf{p}}(\mathbf{x})d^2\mathbf{x}, \quad \mathbf{p} \in K \tag{Ap4.15}$$

Instead of the coefficients $w_{\mathbf{p}}$, Eqs.(Ap4.4), (Ap4.10), and (Ap4.11). In this case, the error of the approximation of the field by the truncated series

$$\varepsilon^2(L) = <\int_G \left| w(\mathbf{x}) - \sum_{\mathbf{p}\in K} \hat{w}_{\mathbf{p}} \Psi_{\mathbf{p}}(\mathbf{x}) \right|^2 d^2\mathbf{x}> \tag{Ap4.16}$$

increases as compared with the minimum value of $\varepsilon_0^2(L)$, Eq.(Ap4.13). The coefficients $\hat{w}_{\mathbf{p}}$ attaining small correlations characterized by the matrix of correlations

$$R_{\mathbf{pp}} = < \hat{w}_{\mathbf{p}} \hat{w}_{\mathbf{p}}{}^* > \tag{Ap4.17}$$

and by the root-mean-square deviation from the diagonal correlation matrix of the ideal canonical basis:

$$\varepsilon_c(L) = \left( \frac{1}{L^2} \sum_{\mathbf{p}\in K} \sum_{\mathbf{p}'\in K} \left| \hat{R}_{\mathbf{pp}'} - \lambda_{\mathbf{p}} \delta_{\mathbf{pp}'} \right|^2 \right)^{1/2} . \tag{Ap4.18}$$

Let us estimate $\varepsilon^2(L)$ and $\varepsilon_c(L)$. We shall substantially use the property of decorrelation of the canonical basis coefficients, Eq.(Ap4.11) for performing estimations.

To derive the estimate of the value $\varepsilon^2(L)$, based of the orthogonality of the function $\{\Psi_{\mathbf{p}}\}$, we can write

$$\varepsilon^2(L) = \varepsilon_0^2(L) + \sum_{\mathbf{p}\in K} < \left| \hat{w}_{\mathbf{p}} - w_{\mathbf{p}} \right|^2 >$$

Owing to Eq.(4.10), we find that

$$\hat{w}_{\mathbf{p}} = w_{\mathbf{p}} + \sum_{\mathbf{p}'} H^{(\mathbf{p})}_{\mathbf{pp}'} w_{\mathbf{p}'}, \tag{Ap4.19}$$

Therefore, if we take into account Eq.(Ap4.11) and the property of eigenvalues $\{\lambda_{\mathbf{p}}\}$ of being arranged in the decreasing order, we obtain

$$\varepsilon^2(L) \le \varepsilon_0^2(L) + \lambda_0 \sum_{\mathbf{p}\in K} \sum_{\mathbf{p}'\in K} \left| H^{(\mathbf{p})}_{\mathbf{pp}'} \right|^2 + \bar{\lambda} \sum_{\mathbf{p}\in K} \sum_{\mathbf{p}'\notin K} \left| H^{(\mathbf{p})}_{\mathbf{pp}'} \right|, \tag{Ap4.20}$$

where

$$\bar{\lambda} = \max_{\mathbf{p}\notin K} \lambda_{\mathbf{p}}, \quad \lambda_0 = \max_{\mathbf{p}} \lambda_{\mathbf{p}} \tag{Ap4.21}$$

Hereinafter, it seems convenient to use, the concept of the absolute norm $a^2(\cdot)$ (see [Ahiezer, 1966]) for the finite or infinite matrix as a sum of squares of the matrix elements' modules.

It is known that the absolute norm of the sum or product of matrices do not exceed the sum or the product of their absolute norms respectively. The estimate for $\varepsilon^2(L)$, Eq.(4.33), can now be rewritten as follows

$$\varepsilon^2(L) \le \varepsilon_0^2(L) + \lambda_0 a^2(H_L) + \bar{\lambda}[a^2(H) - a^2(H_L)] \qquad \text{(Ap4.22)}$$

where $H_L$ matrix have been defined by Eq.(4.19),

$$a^2(H_L) = \sum_{\mathbf{p},\mathbf{p}\in K} \left| H_{\mathbf{pp'}}^{(\mathbf{p})} \right|^2 .$$

$H$ is the infinite matrix comprising all the matrix elements $H_{\mathbf{pp'}}^{(\mathbf{p})}$ Eq.(4.7). By virtue of the Parseval equation, Eq.(4.16), we can write the absolute norm of the $H$ matrix in the form

$$a^2(H) = \sum_{\mathbf{p}\in K} \left\| h_{\mathbf{p}} \right\|^2, \quad \left\| h_{\mathbf{p}} \right\|^2 = Q^{(\mathbf{p})}. \qquad \text{(Ap4.23)}$$

Let us carry out the estimation of the correlation (4.18). From Eqs.(Ap4.19), (Ap4.11), and (Ap4.17) we have that

$$R_{\mathbf{pp'}} = \lambda_{\mathbf{p}} \delta_{\mathbf{pp'}} + \lambda_{\mathbf{p'}} H_{\mathbf{pp'}}^{(\mathbf{p})} + \lambda_{\mathbf{p}} H_{\mathbf{p'p}}^{(\mathbf{p'})^*} + \sum_{\mathbf{p''}} \lambda_{\mathbf{p''}} H_{\mathbf{pp''}}^{(\mathbf{p})} H_{\mathbf{p'p''}}^{(\mathbf{p'})^*}$$

Allowing for the definition and properties of the absolute norm (see [Ahiezer, 1966]) we can write

$$\varepsilon_c(L) = \frac{1}{L} a([R_{\mathbf{pp'}} - \lambda_{\mathbf{p}} \delta_{\mathbf{pp'}}]) \qquad \text{(Ap4.24)}$$

therefore,

$$\varepsilon_c(L) \le \frac{1}{L} \{2\lambda_0 a(H_L) + \lambda_0 a^2(H_L) + \bar{\lambda}[a^2(H) - a^2(H_L)]\} \qquad \text{(Ap4.25)}$$

Rougher estimation use the following inequalities:

$$\lambda_0 \le \sqrt{\sum_{\mathbf{p}\in K} \lambda_{\mathbf{p}}^2} \text{ and } \bar{\lambda} \le \sqrt{\int\int |R(\mathbf{x},\mathbf{x'})|^2 d^2\mathbf{x}\, d^2\mathbf{x'} - \sum_{\mathbf{p}\in K} \lambda_{\mathbf{p}}^2}$$

It has been shown in [Golub, 1980] that the feasibility arises to decrease $\varepsilon^2$ and $\varepsilon_c$ if the perturbation correction is performed by means of "mixing" the $w_{\mathbf{p}}$ coefficients using the following equation

$$(w_{\mathbf{p}}^{(r)}) = \sum_{r'=0}^{r} (-H_L^*)^{r'} (w_{\mathbf{p}}) \qquad \text{(Ap4.26)}$$

The value $r$ is called the order of the correlation.

Let us examine how the error $\varepsilon^2$ depends on $L$ number. With the increasing $L$ the value $\varepsilon_0^2(L)$ decreases, whereas the value $\varepsilon_h^2(L)$ increases. Therefore, the total error should have its minimum in some point $L = L_0$, with s determined by the perturbations of the base-functions. The minimum attainable value of the total error is not equal to zero as is the case for the ideal basis, but equals $\varepsilon^2(L_0)$. There is no reason in increasing the number of the terms in excess of $L_0$. Using the foregoing estimations, Eqs.(Ap4.22) and (Ap4.23), we can obtain the quantitative value of the limit $L_0$.

As an example of perturbations of base functions, Eqs.(Ap2.1)-(Ap2.3), let us consider the Karhunen-Loeve expansions with the biexponential correlation function (Ap. 4.14), at ($\beta_1$=$\beta_2$=0). With respect to the factorizable sinusoidal base-functions, Eq.(Ap2.1), the perturbation matrix can be calculated exactly. Let us make use of Eqs.(Ap2.4) and (Ap2.7) of Appendix 2. By substituting there the proper values of $E_{xp}$ and $E_{yl}$ we find that

$$H_{plp'l'}^{(p,l)} = H_{pp'}(\tau_1, N_1)H_{ll'}(\tau_2, N_2) + H_{pp'}(\tau_1, N_1)\delta_{ll'} + \delta_{pp'}H_{ll'}(\tau_2, N_2) \quad \text{(Ap4.27)}$$

For $(p \neq p')$ we have:

$$H_{pp'}(\tau, N) = \frac{1}{\sqrt{1+e_p(\tau)}} \cdot \frac{1}{\sqrt{1+e_p(\tau)}} \cdot \frac{\sin[z_p(\tau)/N]}{z_p(\tau)} \cdot$$

$$\cdot \left\{ \frac{\sin[z_p(\tau) - z_{p'}(\tau)]}{\sin\left[\dfrac{z_p(\tau) - z_{p'}(\tau)}{N}\right]} \cos(\alpha_p - \alpha_{p'}) - \frac{\sin[z_p(\tau) + z_{p'}(\tau)]}{\sin\left[\dfrac{z_p(\tau) + z_{p'}(\tau)}{N}\right]} \cos(\alpha_p + \alpha_{p'})) \right\}$$

(Ap4.28)

$$H_{pp}(\tau, N) = \frac{1}{1+e_p(\tau)} \cdot \frac{\sin[z_p(\tau)/N]}{z_p(\tau)} \cdot \left\{ N + (-1)^p \frac{\sin[z_p(\tau)]}{\sin[\dfrac{2z_p(\tau)}{N}]} \right\} - 1, \quad \text{(Ap4.29)}$$

where the values $z_p$, $l_p$, and $\alpha_p$ are specified by the equations of section 4.1 of the present Appendix.

Now based on the definition, the value $a^2(H)$ is readily determined. Using Eqs.(Ap4.23), (4.16), and (4.54). Calculating the norms of the derivatives gives us the final equation

$$a^2(H) = \frac{1}{3N_1^2} \sum_{(p,l)\in K} z_p^2(\tau_1) \frac{1-e_p(\tau_1)}{1+e_p(\tau_1)} + \frac{1}{3N_2^2} \sum_{(p,l)\in K} z_l^2(\tau_2) \frac{1-e_l(\tau_2)}{1+e_l(\tau_2)}, \quad \text{(Ap4.30)}$$

(see Fig. Ap4.3). Now, all the data necessary to calculate Eqs.(Ap4.22) and (Ap4.25) is available. Table Ap4.1 summarizes the error taking place when using computer-generated spatial filters with various numbers of pixels $N = N_1 = N_2$.

The curves in Figs. Ap4.4 and Ap4.5 serve to illustrate our general statement about the existence of the minimum of the curve $\varepsilon^2(L)$. The values of the limit $L_0$ for some $N = N_1 = N_2$ are presented in Table Ap4.2.

Note that the point of minimum $L_0$ shifts to the right with the increasing number N of pixels, that is, with the decreasing perturbations. This means that a decrease in perturbations results in the improvement of both the finite-dimensional approximation (Ap.4.16), for every L, and the best-achievable accuracy $\varepsilon^2(L_0)$.

Figure Ap4.5 shows that in accordance with the theoretical estimations an increase of the correction order $r$ yields the displacement of the curves downwards, with their minimum points to be displaced along the L axis to the right. The plot for the accuracy $\varepsilon_0^2(L) = \varepsilon^2(L,\infty)$ best achievable for $r \to \infty$ limits the lowering of the curves. The dashed curve depicts the function $\varepsilon_0^2(L)$ corresponding to the non-perturbed basis. Figures Ap.4.6 and Ap.4.7 show that the $\varepsilon_c$ correlations increase monotonically with the increasing L and they can be diminished thanks to an increase in the number N of pixels or due to conducting the correction, Eq.(Ap4.26).

**B. Perturbations of the correlation function**

Another type of perturbations is related to the perturbations of the kernel of Eqs.(Ap4.10) and (Ap4.1) that determines the basis functions for the Karhunen-Loeve expansion.

In a variety of cases, we have to determine the basis functions for the Karhunen-Loeve expansion of a random field $w(\mathbf{x}), \mathbf{x} \in G$ on the basis of the kernel $\tilde{R}(\mathbf{x},\mathbf{x}')$ that is only approximation of the correlation function $R(\mathbf{x},\mathbf{x}')$.

In this case, solving an integral equation results in obtaining the eigenfunctions $\{\tilde{\Psi}_\mathrm{p}(\mathbf{x})\}$ and eigenvalues $\{\tilde{\lambda}_\mathrm{p}\}$ of the kernel $\tilde{R}(\mathbf{x},\mathbf{x}')$, in general terms being different from the true eigenfunctions $\{\psi_\mathrm{p}\}$ and eigenvalues $\{\lambda_\mathrm{p}\}$ of the kernel $R(\mathbf{x},\mathbf{x}')$. We should specify the conditions under which the approximation

$$\tilde{R}(\mathbf{x},\mathbf{x}') \cong R(\mathbf{x},\mathbf{x}')$$

is matched to the approximation

$$\tilde{\Psi}_\mathrm{p}(\mathbf{x}) \cong \Psi_\mathrm{p}(\mathbf{x}), \quad \tilde{\lambda}_\mathrm{p} \cong \lambda_\mathrm{p}.$$

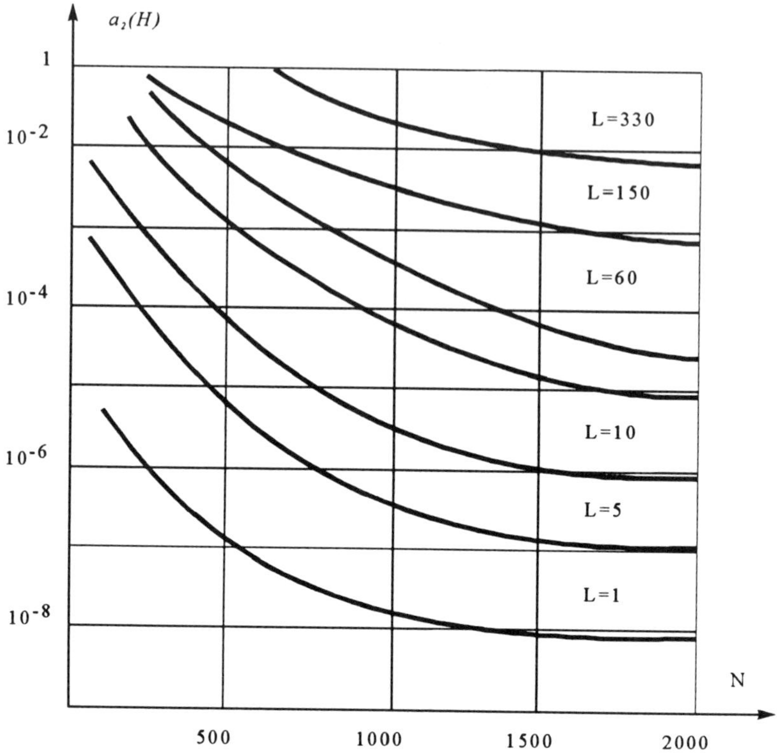


**Figure Ap.4.3.** The total error of the sampling of the Karhunen-Loeve base functions ($\tau_1 = \tau_2 = 2,3$). (Modified from Golub M.A., Soifer V.A., The stability of Karhunen-Loeve expansion and computer generation of optimal spatial filters, in *Spectral methods of information processing in scientific research*, Center for biological research, Pushino, 1980, 108 (in Russian)).

**Table Ap4.1**

| N | 128 | 356 | 512 | 1024 |
|---|---|---|---|---|
| $\varepsilon_c/\lambda_0$ | $5.970\,10^{-3}$ | $1.260\,10^{-3}$ | $0.303\,10^{-3}$ | $0.075\,10^{-3}$ |
| $\varepsilon^2/E$ | 0.0989 | 0.0577 | 0.0548 | 0.0546 |

The corresponding task relates to the theory of perturbations covered by quantum mechanics (see [Landau, 1974]). Let us call the kernel $R(\mathbf{x},\mathbf{x}')$ non-perturbed, the kernel $\tilde{R}(\mathbf{x},\mathbf{x}')$ - perturbed, and the kernel $\Delta R(\mathbf{x},\mathbf{x}') \cong \tilde{R}(\mathbf{x},\mathbf{x}') - R(\mathbf{x},\mathbf{x}')$, $\mathbf{x},\mathbf{x}' \in G$ - the kernel's perturbation. Let us examine the case when the eigenvalues of the non-perturbed kernel are 1-multiple only. In this case from the theory of the first-order of perturbations we find that [Landau, 1974]

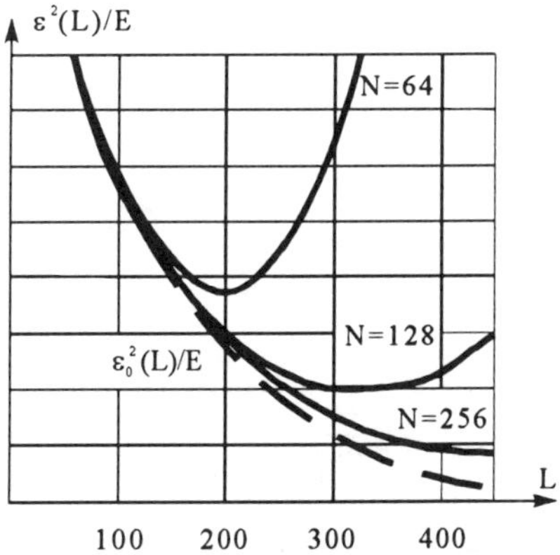


**Figure Ap.4.4.** Mean-square deviation in the case of the perturbed base-functions ($\tau_1=\tau_2=2.3$). (Modified from Golub M.A., Soifer V.A., The stability of Karhunen-Loeve expansion and computer generation of optimal spatial filters, in *Spectral methods of information processing in scientific research*, Center for biological research, Pushino, 1980, 108 (in Russian)).

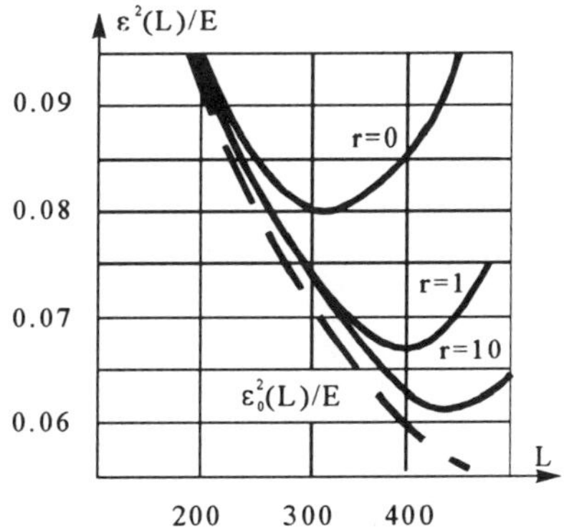


**Figure Ap.4.5.** Mean-square deviation after conducting the correction ($\tau_1=\tau_2=2,3$). (Modified from Golub M.A., Soifer V.A., The stability of Karhunen-Loeve expansion and computer generation of optimal spatial filters, in *Spectral methods of information processing in scientific research*, Center for biological research, Pushino, 1980, 108 (in Russian)).

**Table Ap4.2**

| | | | |
|---|---|---|---|
| N | 64 | 128 | 256 |
| $L_0$ | 184 | 320 | 464 |
| $\varepsilon^2(L_0)/E$ | 0.120 | 0.080 | 0.057 |

$$\tilde{\lambda}_{\mathbf{p}} = \lambda_{\mathbf{p}} + \Delta R_{\mathbf{pp}} \tag{Ap4.31}$$

$$\tilde{\Psi}_{\mathbf{p}}(\mathbf{x}) = \Psi_{\mathbf{p}}(\mathbf{x}) + \sum_{\mathbf{p'}\neq\mathbf{p}} \frac{\Delta R_{\mathbf{p'p}}}{\lambda_{\mathbf{p}} - \lambda_{\mathbf{p'}}} \Psi_{\mathbf{p'}}(\mathbf{x}), \tag{Ap4.32}$$

where

$$\Delta R_{\mathbf{p'p}} = \int\int_{G\times G} \Delta R(\mathbf{x},\mathbf{x}')\, \Psi^*_{\mathbf{p'}}(\mathbf{x}) \psi_{\mathbf{p}}(\mathbf{x}')\, d^2\mathbf{x}\; d^2\mathbf{x}' \tag{Ap4.33}$$

are the matrix elements of the kernel's perturbation $\Delta R(\mathbf{x},\mathbf{x}')$.

Thus, instead of the base-functions $\{\Psi_{\mathbf{p}}(x),\ \mathbf{p}\in K\}$ we make use of the other basis functions $\{\tilde{\Psi}_{\mathbf{p}}(x),\ \mathbf{p}\in K\}$ and instead of the coefficients $\{w_{\mathbf{p}},\mathbf{p}\in K\}$, Eq.(Ap4.4), we compose the other coefficients $\{\tilde{w}_{\mathbf{p}},\mathbf{p}\in K\}$ in the form

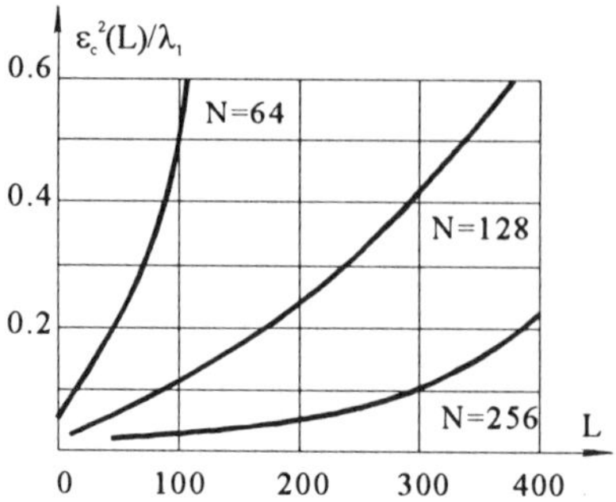


**Figure Ap.4.6.** Correlations between the perturbed coefficients of Karhunen-Loeve expansion ($\tau_1$=$\tau_2$=2.3). (Modified from Golub M.A., Soifer V.A., The stability of Karhunen-Loeve expansion and computer generation of optimal spatial filters, in *Spectral methods of information processing in scientific research*, Center for biological research, Pushino, 1980, 108 (in Russian)).

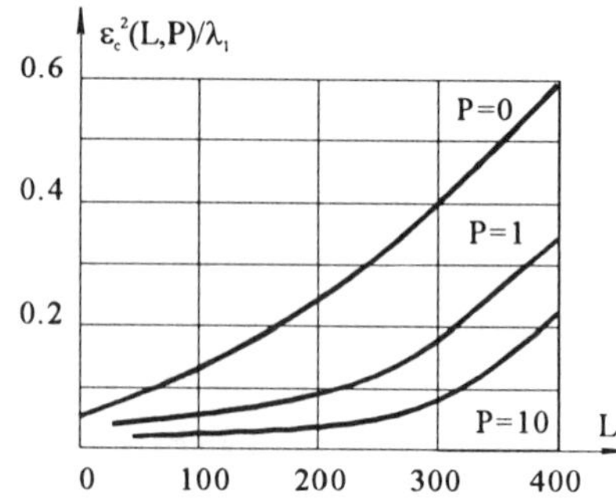


**Figure Ap.4.7.** Correlations between the corrected coefficients of Karhunen-Loeve expansion ($\tau_1$=$\tau_2$=2.3). (Modified from Golub M.A., Soifer V.A., The stability of Karhunen-Loeve expansion and computer generation of optimal spatial filters, in *Spectral methods of information processing in scientific research*, Center for biological research, Pushino, 1980, 108 (in Russian)).

$$\tilde{w}_{\mathbf{p}} = (w, \tilde{\Psi}_{\mathbf{p}}) = w_{\mathbf{p}} + \sum_{\mathbf{p}'\neq\mathbf{p}} \frac{\Delta R^*_{\mathbf{p}'\mathbf{p}}}{\lambda_{\mathbf{p}} - \lambda_{\mathbf{p}'}} \cdot w_{\mathbf{p}'}. \tag{Ap4.34}$$

Therefore, we can see that the perturbations $\Delta R$ of the correlation function $R$ lead to the perturbations

$$s_{\mathbf{p}}(\mathbf{x}) = \tilde{\Psi}_{\mathbf{p}}(\mathbf{x}) - \Psi_{\mathbf{p}}(\mathbf{x}) \tag{Ap4.35}$$

of the basis functions $\{\Psi_{\mathbf{p}}(x),\ \mathbf{p} \in K\}$. The coefficients of the expansion are attaining the small correlations

$$\rho_{\mathbf{pp}'} = < (\tilde{w}_{\mathbf{p}} - w_{\mathbf{p}})(\tilde{w}_{\mathbf{p}} - w_{\mathbf{p}})^* > \tag{Ap4.36}$$

Let us introduce characteristics for estimating the effect of perturbations

$$\varepsilon_\Psi = \left( \frac{1}{L} \sum_{\mathbf{p}\in K} \left\| s_{\mathbf{p}}(\mathbf{x}) \right\|^2 \right)^{1/2} \tag{Ap4.37}$$

$$\varepsilon_\rho = \left(\frac{1}{L^2} \sum_{\mathbf{p}\in K} \sum_{\mathbf{p}'\in K} \left|\rho_{\mathbf{pp}'}\right|^2\right)^{1/2} \tag{Ap4.38}$$

On the basis of Eqs.(Ap4.31), (Ap4.32), (Ap4.3), and (Ap4.1) we find that

$$\rho_{pp'} = \sum_{p''\neq p,p'} \frac{\lambda_{p''}\Delta R^*_{p''p}\Delta R_{p''p'}}{(\lambda_p - \lambda_{p''})(\lambda_{p'} - \lambda_{p''})} \quad \text{(Ap4.39)}$$

$$\|s_p\|^2 = \sum_{p'\neq p} \frac{|\Delta R_{p'p}|^2}{|\lambda_p - \lambda_{p'}|^2} \quad \text{(Ap4.40)}$$

We introduce the values

$$\Delta\lambda(L) = \min_{p\in K} \Delta\lambda_p, \quad \Delta\lambda_p = \min_{p'\neq p} |\lambda_p - \lambda_{p'}|. \quad \text{(Ap4.41)}$$

From Eqs.(Ap4.39) and (Ap4.40) we obtain the following estimations

$$|\rho_{pp'}| \le \frac{\lambda_0}{\Delta\lambda_p \Delta\lambda_{p'}} \left( \sum_{p''\neq p'} |\Delta R_{p''p}|^2 \right)^{1/2} \left( \sum_{p''\neq p} |\Delta R_{p''p'}|^2 \right)^{1/2}$$

$$\|s_p\|^2 = \frac{1}{(\Delta\lambda_p)^2} \sum_{p'\neq p} |\Delta R_{p'p}|^2$$

from which it follows that

$$\varepsilon_\Psi \le \frac{1}{\sqrt{L}\,\Delta\lambda(L)} \left[ a^2(\Delta R) - \sum_{p\in K} (\Delta R_{pp})^2 \right]^{1/2}, \varepsilon_\rho \le \lambda_0 \varepsilon_\psi^2 \quad \text{(Ap4.42)}$$

where

$$a^2(\Delta R) \equiv \sum_p \sum_{p'} |\Delta R_{pp'}|^2 = \iint_{G\times G} |\Delta R(\mathbf{x},\mathbf{x}')|^2 d^2\mathbf{x}\, d^2\mathbf{x}' \quad \text{(Ap4.43)}$$

is the absolute norm of the infinite matrix $[\Delta R_{pp'}]$. Thus, all the criteria are expressed through the value $a^2(\Delta R)$.

## REFERENCES


**Adzhalov V.I., Golub M.A., Karpeev S.V., Sisakyan I.N., Soifer V.A.**, Multichannel computer optics components matched to mode groups, *Soviet Journal of Quantum Electronics*, 20(2), 136, 1990.

**Ahiezer N.I., Glazman I.M.**, Theory of linear operators in Hilbert's space, Nauka Publishers, Moscow, 1966 (in Russian).

**Akaev A.A., Maiorov C.A.**, Coherent optical computers, Mashinostroyenie, Leningrad, 1977 (in Russian).

**Anan'ev Yu.A.**, Optical resonators and laser beams, Nauka Publishers, Moscow, 1989 (in Russian).

**Arefyev E.Yu., Bambulevich K.E., Golub M.A., Karpeev S.V., Sisakyan I.N., Soifer V.A.**, Comparison of two methods for the determination of the power distribution between modes, *Soviet Journal of Quantum Electronics*, 18(12), 1548, 1988.

**Arefyev E.Yu., Golub M.A., Ovchinnikov K.V., Popov S.B., Sisakyan I.N., Soifer V.A., Tikhonov D.N., Khramov A.G., Shamalova G.V.**, Monitoring the phase microrelief of computer-optics components, *Soviet Physics - Technical Physics*, 35(6), 732, 1990.

**Baranova N.B., Zeldovich B.Ya.**, Investigation of dislocation density for the wavefront of light fields with speckle structure, *Journal of Experimental and Theoretical Physics*, 5(11), 1702, 1982 (in Russian).

**Bartelt H.O., Case S.K.**, High-efficiency hybrid computer-generated holograms, *Applied Optics*, 21(16), 2886, 1982.

**Bartelt H.O., Lohmann A.W.,Freude W.,Grau G.K.**, Mode analysis of optical fibers using computer-generated matched filters, *Electronic letters*, 19(7), 247, 1983.

**Bartelt H.O.**, Applications of the tandem component: an element with optimum light efficiency, *Applied Optics*, 24(22), 3811, 1985.

**Berdague S., Facq P.**, Mode division multiplexing in optical fibers, *Applied Optics*, 21(11), 1950, 1982.

**Bereznyi A.E., Prokhorov A.M., Sisakyan I.N., Soifer V.A.**, Bessel optics, *Soviet Physics: Doklady*, 29(2), 115, 1984.

**Bereznyi A.E., Karpeev S.V., Uspleniev G.V.**, Computer generated holographic optical elements produced by photolithography, *Optics and Lasers in Engineering*, 15(5), 331, 1991.

**Bergeron A., Arsenault H.H., Gauvin J., Gingras D.J.** Computer-generated holograms improved by global iterative coding. *Optical Engineering*, 32(9), 2216, 1993.

**Brown B.R., Lohmann A.W.**, Complex spatial filtering with binary masks, *Applied Optics*, 5(6), 967, 1966.

**Born M., Wolf E.**, Principles of optics, Pergamon Press, Oxford, London, 1980.

**Campos J., Turon F., Yzuel M.J.**, Pattern recognition with quantized computer-generated filters. *Applied Optics*, 31(17), 3337, 1993.

**Chu D.C., Goodman J.W.,** Spectrum shaping with parity sequences, *Applied Optics*, 11(8), 1716, 1972.

**Chu D.C., Fienup J.R., Goodman J.W.,** Multiemulsion on-axis computer generated holograms, *Applied Optics*, 12(4), 1386, 1973.

**Chu D.C., Fienup J.R.,** Recent approaches to computer-generated holograms, *Optical Engineering*, 13(3), 189, 1974.

**Cozannet A., Fleuret J., Maitre H., Rousseau M.,** Optique et telecommunications, Eyrolles et Cnet-Enst, Paris, 1981.

**Dallas W.J.,** Computer-generated holograms, in *The computer in Optical Research, Topics in Applied Physics*, vol. 41, Frieden B.R. ed., Springer Verlag, Berlin, 1986, p.291.

**Daugavet I.K.,** Introduction to the theory of approximations of functions, State University, Leningrad, 1977 (In Russian).

**Denisyuk Yu.N., Davydova I.N.,** Some particularities of orthogonal function-images registration in a 3-D holography, *Optics and Spectroscopy*, 60(2), 365, 1986, (in Russian).

**Diado Y., Miyauchi E., Twama T., Otsuka T.,** Determination of modal power distribution in graded-index optical wave-guides from near-field patterns and its application to differential mode attenuation measurement, *Applied Optics*, 18(13), 2207, 1979.

**Doskolovich L.L., Kazansky N.L., Kharitonov S.I., Uspleniev G.V.,** Focusators for laser-branding. *Optics and Lasers in Engineering*, 15(5), 311, 1991.

**Egalon C.O., Rogowski R.S.,** Exact solution for phase shift of TE and TM modes in an optical fiber due to axial strain. *Optical Engineering*, 32(7), 1638, 1993.

**Erdely A.,** Higher transcendental functions, vol. 2, McGraw-Hill, New-York, 1953.

**Felsen L.B., Marawitz M.,** Radiation and scattering of waves, Prentice-Hall, Englewood Cliffs, New Jersey, 1973.

**de Fornel F., Arnaud J., Facq P.,** Microbending effects on monomode light propagation in multimode fibers, *Journ. Opt. Soc. Amer*, 73(5), 661, 1983.

**Fu K.S.,** Sequential methods in pattern recognition and machine learning, Academic Press, New York, 1968.

**Gadliardi R.M., Karp S.,** Optical communications, John Wiley and Sons, New York, 1976.

**Gamble F.T., Laferty D.R.,** Hybrid multiplexing matched spatial filters in the fourier plane for multiple object recognition, *Appl. Opt.*, 29(5), 709, 1990.

**Garitchev V.P., Golub M.A., Karpeev S.V., Krivoshlykov S.G., Petrov N.I., Sisakyan I.N., Soifer V.A., Haubenreisser W., Jahn J.-U., Willsh R.,** Experimental investigation of mode coupling in a multimode graded-index fiber caused by periodic microbends using computer-generated spatial filters, *Optics Communications*, 55(6), 403, 1985.

**Garitchev V.P., Golub M.A., Karpeev S.V., Krivoshlykov S.G., Sisakyan I.N., Soifer V.A., Uvarov G.V.,** Use of synthesized holograms for selective mode excitation in gradient index fibers, Analysis of sensitivity to radial displacement of launch beams, in *Computer Optics*, Pergamon Press, London, 2(1), 95, 1990.

**Garitchev V.P., Golub M.A., Karpeev S.V., Krivoshlykov S.G., Sisakyan I.N., Soifer V.A., Uvarov G.V.**, Use of synthesized holograms for selective mode excitation in gradient index fibers, Analysis of sensitivity to radial displacement of launch beams, in *Computer Optics*, Pergamon Press, London, 2(1), 95, 1990.

**Gelfand I.M.**, Lectures in linear algebra, Nauka Publishers, Moscow, 1971 (in Russian).

**Gibin I.S., Nezhevenko E.S., Potaturkin O.I., Tverdokhleb P.E.**, Coherent optical devices for generalized spectral analysis, *Avtometriya*, (5), 3, 1972 (in Russian).

**Golub M.A., Soifer V.A.**, The stability of Karhunen-Loeve expansion and computer generation of optimal spatial filters, in *Spectral methods of information processing in scientific research*, Center for biological research, Pushino, 1980, 108 (in Russian).

**Golub M.A., Prokhorov A.M., Sisakyan I.N., Soifer V.A.**, Synthesis of spatial filters for investigation of the transverse mode composition of coherent radiation, *Soviet Journal of Quantum Electronics*, 12(9), 1208, 1982.

**Golub M.A., Karpeev S.V., Krivoshlykov S.G., Prokhorov A.M., Sisakyan I.N., Soifer V.A.**, Experimental investigation of spatial filters separating transverse modes of optical fields, *Soviet Journal of Quantum Electronics*, 13(8), 1123, 1983.

**Golub M.A., Karpeev S.V., Krivoshlykov S.G., Prokhorov A.M., Sisakyan I.N., Soifer V.A.**, Spatial filter investigation of the distribution of power between transverse modes in a fiber waveguide, *Soviet Journal of Quantum Electronics*, 14(9), 1255, 1984.

**Golub M.A., Karpeev S.V., Kazanskyi N.L., Mirzov A.N., Sisakyan I.N., Soifer V.A., Uvarov G.V.**, Spatial phase filters matched to transverse modes, *Soviet Journal of Quantum Electronics*, 18(3), 392, 1988.

**Golub M.A., Kazanskiy N.L., Prokhorov A.M., Sisakyan I.N., Soifer V.A.**, Synthesis of optical antennae, *Computer Optics*, Pergamon Press, London, 1(1), 25, 1989a.

**Golub M.A., Sisakyan I.N., Soifer V.A., Uvarov G.V.**, Optical components for analysis and formation of the transverse mode composition, *Soviet Journal of Quantum Electronics*, 19(4), 543, 1989b.

**Golub M.A.**, Estimate of the accuracy of formation of an aspherical wavefront, *Optics and Spectroscopy*, 68(2), 226, 1990a.

**Golub M.A.**, Quantization in standard mode-selecting elements of computer-synthesized optics, *Computer Optics*, Pergamon Press, London, 2(1), 29, 1990b.

**Golub M.A., Sisakyan I.N., Soifer V.A.**, Computer generated optical elements for fiber's mode selection and launching, *Proceedings SPIE, OE/Fibers'90*, 1365, 156, 1990c.

**Golub M.A., Sisakyan I.N., Soifer V.A., Uvarov G.V.**, New measurement techniques for modal power distribution in fibers, *Proceedings SPIE, OE/Fibers'90*, 1366, 273, 1990d.

**Golub M.A., Sisakyan I.N., Soifer V.A.**, Phase quantization and discretization in diffractive optics, *Proceedings SPIE, San-Diego'90*, 1334, 188, 1990e.

**Golub M.A., Sisakyan I.N., Soifer V.A.,** Computer generated optical elements in wavefront formation with intensity spatial modulation, *Modern Optics*, 38(6), 1067, 1991a.

**Golub M.A, Sisakyan I.N., Soifer V.A.,** Mode selection of laser radiation by computer-generated optical elements, *Optics and lasers in Enginereeng*, 15(5), 314, 1991b.

**Golub M.A., Sisakyan I.N., Soifer V.A., Uvarov G.V.,** Mode-selective fiber sensors operating with computer generated optical elements, *Proceedings SPIE, OFSC'91*, 1572, 101, 1991c.

**Golub M.A.** Generalized complex-to-phase coding for computer- generated holograms, in *Proceedings of the Third Int. Seminar on Digital image processing in medicine, remote sensing and visualisation of information,* Riga, p.77, 1992a .

**Golub M.A., Doskolovich L.L., Kazansky N.L., Kharitonov S.I., Soifer V.A.,** Computer generated diffractive multi-focal lens, *Journal of Modern Optics*, 39(6), 1245, 1992b.

**Goodman J.W.,** Introduction to Fourier Optics, McGraw-Hill, New-York, 1968.

**Gowar J.,** Optical communication systems, Prentice Hall International Inc., London, 1984.

**Grau G.K., Leminger O.G., Sauter E.G.,** Mode excitation in parabolic index fibers by Gaussian beams, *Archiv fur Electronik und Ubertragungstechnik, Electronics and Communication*, 34, 259, 1980.

**Haus H.A.,** Waves and Fields in Optoelectronics, Prentice Hall, Inc. Englewood Cliffs, 1984.

**Huang T.S.,** Digital Holography, Proc. IEEE, 59(9), 1335, 1971.

**Janke E., Emde F., Lösch F.,** Tafeln Höherer Funktionen, B.G. Teubner Verlagsgesellschaft, Stuttgart, 1960.

**Kachmazh S., Shteingauz G.,** Theory of orthogonal expansions, Nauka Publishers, Moscow, 1958 (in Russian).

**Karpeev S.V, Uvarov G.V.,** Automatic adjustment and control of mode spatial filters, in *Optical recording and processing of information*, Soifer V.A., Ed., Kuibyshev Aviation Institute, 1988, p.36 (in Russian).

**Kashin V.A.,** Synthesis of antenna arrays by the method fast-oscillating phase, *Radiotehnika i Elektronika*, 15(1), 170, 1970, (in Russian).

**Kapany N.S.,** Fiber Optics; principles and application, Academic, New York, 1967.

**Karhunen K.,** Uber linear Methoden in der Wahrscheinlichkeitsrechnung, *Ann. Acad. Sci. Fennicae*, Helsinki, 1947, Ser. A137.

**Kirk J.P., Jones A.L.,** Phase-only complex-valued spatial filter, Journ. *Opt. Soc. Amer.*, 61(8), 1023, 1971.

**Kitayama K.I., Takeda M., Seikai S., Vchida N.,** Determination of mode power distribution in parabolic index optical fibers, *IEEE J. Quantum El.*, QE-15(10), 1161, 1979.

**Klovskiy D.D. , Soifer V.A.**, Space-and-time signal processing, Svyaz Publishers, Moscow, 1976, (in Russian).

**Korn G.A., Korn T.A.**, Mathematical Handbook, McGraw-Hill Book Company, 1968.

**Kotlyar V.V.**, Decomposition of coherent field by orthogonal basis, *Computer Optics*, 5, 31, 1989 (in Russian).

**Krackhardt U., Schwider J., Schrader M., Streibl N.**, Synthetic holograms written by a laser pattern generator. *Optical Engineering*, 32(4), 781, 1993.

**Krivoshlykov S.G., Petrov N.I., Sisakyan I.N.**, Modal energy transformation between two connected multi-mode square-law index optical waveguides, *Opt. and Quant. Electr.*, 15, 193, 1983.

**Krivoshlykov S.G., Sauter E.**, Mode coupling between two waveguides with offset, tilt and gap, using quantum theoretical methods, *Journal of Physics*, A20, 3805, 1987.

**Kuvshynov A.P., Vorontzov B.U.**, Helmholtz equation solutions for wave beam, *Ukrainski physich. zhurnal*, 33(11), 1658, 1988 (in Russian).

**Landau L., Lifchitz E.**, Quantum mechanics, Nauka Publishers, Moscow, 1974, (in Russian).

**Landesman B.T., Barrett H.H.**, Gaussian amplitude functions that are exact solutions to the scalar Helmholtz equation, *J. Opt. Soc. Am.*, A5(10), 1610, 1988.

**Lee S.H.**, Mathematical operations by optical processing, Opt. Eng., 13(3), 196, 1974.

**Lee S.H.**, Diffractive optics and computer-generated holograms for optical interconnects, *Proceedings SPIE, San-Diego'93*, CR49, CR49-14, 1993.

., 13(3), 196, 1974.

**Lesem L. B., Hirsch P., Jordan J.A.**, The kinoform: a new wavefront reconstruction device, *IBM J. Res. Develop.*, 13, 150, 1969.

**Lissak B., Ruschin S.**, Transverse pattern modification in a stable apertured laser resonator, *Appl.Opt.*, 29(6), 767, 1990.

**Mait J.N., Brenner K.H**. Dual-phase holograms: improved design, Appl. Opt., 26, 4883, 1987.

**Marcuse D.**, Light transmission optics, Van Nostrand Reinhold, New York, 1982.

**Miler M.**, Holografie, SNTL, Praha, 1974.

**Morrison R.L., Walker S.L., Cloonan T.J.**, Beam array generation and holographic interconnections in a free-space optical switching network. *Applied Optics*, 32(14), 2512, 1993.

**Nemoto S.**, Nonparaxial Gaussian beams, Appl. Opt., 29(13), 1940, 1990.

**Nishi I. , Oguchi T., Kato K.**, Broad passband multi/demultiplexer for multimode fibers using a diffraction grating and retroreflectors, *Journal of Lightwave Technology*, LT-5(12), 2411, 1987.

**Papoulis A.**, Systems and transforms with applications in optics, McGraw-Hill, New York, 1968.

**Perelomov A.M.,** Generalized coherent states and their applications, Springer-Verlag, 1986.

**Pressley R.I., Ed.,** Handbook of lasers with selected data on optical technology, Chemical Rubber Co., Cleveland, 1971.

**Rudin W.,** Functional Analysis, McGraw- Hill, New York, 1973.

**Shigezawa H., Matsuo T., Takiyama K.,** Measurement of excitation condition and quantitative mode analysis in optical fibers, *IEEE Transactions on Microwave Theory and Techniques*, MTT-26(12), 992, 1978.

**Sisakyan I.N., Soifer V.A.,** Elements of fine optics generated by computer, paper presented at Intern. Conf. and School Lasers and Applications, Bucharest, 853, 1982.

**Sisakyan I.N., Soifer V.A.,** Fine Optics synthesized by a computer, Abstract of papers presented at Fifth Intern. Conf. Lasers and their Applications, Dresden, 23, 1985.

**Sisakyan I.N., Soifer V.A.,** Computer Optics: achievements and problems, *Computer Optics*, Pergamon Press, London, 1(1), 3, 1989.

**Sisakyan I.N., Soifer V.A.,** Modans - Optical Elements for analysis and synthesis of laser mode structure, in *Computer Optics*, 2(2), Pergamon Press, London, 109, 1990.

**Solimeno S., Crosignani B., Di Porto P.,** Guiding, diffraction and confinement of optical radiation, Academic Press, Inc., 1986.

**Stewart W.J.,** Optical fiber transmission, Williamsburg, Va.: OSA/IEEE, January, 1975.

**Svelto O.,** Principles of Lasers, Plenum Press, New York, 1982.

**Taylor H.F.,** Bending effects in optical fibers, *Journ. of Lightwave Technology*, LT-2(5), 617, 1984.

**Tikhonov A.N., Arsenin V.Y.** Methods for solving ill-posed problems, Nauka Publishers, Moscow, 1986 (in Russian).

**Unger H.G.,** Planar optical waveguides and fibers, Oxford, New York, 1977.

**Vainshtein L.A.,** Opened cavities and opened waveguides, Soviet Radio, Moscow, 1966 (in Russian).

**Vasara A., Taghizadeh M.R., Turunen J., Westerholm J., Noponen E., Ichikawa H., Miller J.M., Jaakkola T., Kuisma S.,** Binary surface-relief gratings for array illumination in digital optics. *Applied Optics*, 31(17), 3320, 1993.

**Walker S.J., Jahns J., Li L., Mansfield W.M., Mulgew P., Tennant D.M., Roberts C.W., Wang L., Geng W.G., Kostuk R.K.,** Efficient multiple-image holographic element, *Opt. Eng.*, 29(3), 257, 1990.

**Watanabe S.,** Karhunen-Loeve expansion and factor analysis, Trans. of the 4th Prague Conf. on Inform. Theory, Publ. House of the Czechoslovak acad. of Sciences, Prague, p.635, 1967.

**West L.C., Ailawadi N.K.,** Design and fabrication of high-efficiency beam splitters and beam deflectors for integrated planar micro-optic systems. *Applied Optics*, 32(14), 2494, 1993.

**Wyrowski F., Bryngdahl O.,** Digital holography as part of diffractive optics. *Reports on Progress in Optics*, 54(12), 1481, 1991.

**Yariv A.,** Optical electronics, Holt, Rinehart, and Winston, New York, 1985.

**Yaroslavskii L.P.,** Applied problems of digital optics, in *Advances in Electronics and Electron Physics*, Vol. 66, Hawkes P.W., Ed., Academic Press, New York, 1986.

**Yaroslavskii L.P., Merzlyakov N.S.,** Methods of digital holography, Plenum Publishing Corporation, New York-London, 1980.

**Youla D.C.,** The solution of a homogeneous Wiener-Hopf integral equation occurring in the expansion of second-order stationary random functions, *IRE Trans. on Inf. Theory*, IT-3(3), 187, 1957.

**Zhang Y.M., Zhou G.,** Multi-focal-length and multichannel holographic optical element, Optics in Complex Systems, paper presented at 15-th Congress of the Int. Commission for Optics, SPIE, vol. 1319, 301, 1990.

# INDEX

## A



## B



## C

**D**

**G**



**H**



**I**

**O**



**P**

**Q**



**R**

**S**



**T**

## U



## W



## Z